CONTROLLING RISK

CONTROLLING
RISK
IN A
DANGEROUS
WORLD

30 Techniques for Operating Excellence

How to Predict and Prevent the Next Accident
and Improve Performance and Productivity in
Hazardous Operations, Business, and Everyday Life

Jim Wetherbee

CAPT. JIM WETHERBEE, USN, RET.

New York

CONTROLLING RISK
IN A DANGEROUS WORLD

30 Techniques for Operating Excellence

How to Predict and Prevent the Next Accident
and Improve Performance and Productivity in
Hazardous Operations, Business, and Everyday Life

© 2017 **JIM WETHERBEE LLC.**

Published in New York, New York, by Morgan James Publishing. Morgan James and The Entrepreneurial Publisher are trademarks of Morgan James, LLC. www.MorganJamesPublishing.com

The Morgan James Speakers Group can bring authors to your live event. For more information or to book an event visit The Morgan James Speakers Group at www.TheMorganJamesSpeakersGroup.com.

Shelfie

A **free** eBook edition is available
with the purchase of this print book.

CLEARLY PRINT YOUR NAME ABOVE IN UPPER CASE

Instructions to claim your free eBook edition:
1. Download the Shelfie app for Android or iOS
2. Write your name in **UPPER CASE** above
3. Use the Shelfie app to submit a photo
4. Download your eBook to any device

ISBN 978-1-63047-950-3 paperback
ISBN 978-1-63047-952-7 eBook
ISBN 978-1-63047-951-0 hardcover
Library of Congress Control Number:
2016900809

Cover Design by:
Kelly Wetherbee and Kevin Smith

Cover Photo Credit:
NASA

Interior Design by:
Bonnie Bushman
The Whole Caboodle Graphic Design

In an effort to support local communities and raise awareness and funds, Morgan James Publishing donates a percentage of all book sales for the life of each book to Habitat for Humanity Peninsula and Greater Williamsburg.

Get involved today, visit
www.MorganJamesBuilds.com

Habitat
for Humanity®
Peninsula and
Greater Williamsburg
Building Partner

In memory of the crews of:

AS-204 (*Apollo 1*)
STS-51-L (*Challenger*)
STS-107 (*Columbia*)

TABLE OF CONTENTS

PREFACE

We live in a dangerous world. There are countless ways to be hurt or to die prematurely in preventable accidents. Since becoming a naval aviator, I have been on a continuous journey of learning how to prevent the next accident that is inevitably trying to injure or kill me.

Every potential accident gives signals before it becomes an accident. To enhance our chance of preventing catastrophe, we must learn to discern these signals. To do this, we study history. We analyze previous organizational failures and catastrophes. Root cause specialists identify problems that, if corrected, create a higher likelihood of preventing future occurrences of similar tragedies. With experience, we learn to prevent accidents.

But can we predict all accidents from observing the past? Are some unpreventable? We easily prevent potential accidents that are similar to recent occurrences, but preventing accidents that exceed corporate experience seems extraordinarily difficult. Organizations continue to be blindsided by tragedies that no one thought would occur. Yet, in any given postincident analysis, investigators often determine the latest catastrophe was tragically similar to a forgotten previous incident. New rules are promulgated, operating procedures are updated—and the cycle of accidents continues. Organizations must need something more than rules and procedures to prevent accidents.

Managers in the organization manage risk with systematic and structured processes intended to limit the assessed risk. But, even in the best organizations, when it is time to go to work, operators don't *manage* risk; they *control* risk. To work effectively and stay alive, the front-line workers need operating techniques for controlling risk to supplement the structured rules and procedures promulgated by managers to manage risk.

This is a book about controlling risk. After almost forty years in hazardous endeavors, I have learned the techniques I use to control risk not only help me stay alive, a fairly nice incentive, but they also help me accomplish more missions in better ways. These are the same techniques necessary for operating excellence, which results in higher performance and greater success for me as an operator, and more profits and maximized long-term productivity for my organization. When really understood and embraced as a way of operating, these techniques enable groups of people working together to optimize results in any high-risk business and accomplish more in our dangerous world—or out of this world.

The human brain is wired for taking risk. Evolution has helped us learn from our ancestors how to enter dangerous situations and, when not succumbing to the hazards, achieve spectacular feats. The result of our wired-plus-acquired risk-taking skills is that we learned how to hunt on prehistoric savannas and fly to the moon from the Florida savanna. We rose to the top of the food chain in this dangerous world because we mastered the skills of taking risks, achieving goals, and staying alive. Throughout evolutionary history, the best stayed alive just long enough to pass along their risk-taking, goal-achieving, alive-staying genes to the next generation.

Many wonderful books have been written about how organizations should manage risk. Two major theories were developed about the relationship between managing risk and organizational performance. Charles Perrow originated one, which he calls *Normal Accident Theory*. He writes that accidents are inevitable because today's systems and the organizations controlling those systems are complex and interdependent, with tight coupling and short time periods between cascading events that lead to accidents.[1] Karl Weick and others have described a seemingly opposing theory when they write about *High Reliability Organizations*.[2] According to this theory, mindful people in resilient organizations can and do prevent accidents. The key is to identify signals of minor problems early and take appropriate actions to prevent escalation to disaster.

The authors of each theory are brilliant writers who have completed extensive, relevant, and valuable research. They have conducted insightful analyses of the data collected and observations made while working with many organizations engaged in

---

1   Charles Perrow, *Normal Accidents: Living with High-Risk Technologies* (Princeton, NJ: Princeton University Press, 1999).

2   Karl E. Weick and Kathleen M. Sutcliffe, *Managing the Unexpected* (San Francisco: Jossey-Bass, 2001).

high-hazard operations. But, if these authors are talented and intelligent academic researchers, who have conducted justifiable analyses, how do we have two theories that seem to conflict?

Why does Normal Accident Theory postulate that accidents are inevitable, while High Reliability Organization theory proposes that accidents can be prevented? From my observations and assessments made while working in dangerous environments, I think *both* theories are correct; both accurately describe operations in hazardous environments.

How can accidents be inevitable and preventable at the same time? Here is how I reconcile this apparent contradiction.

I believe accidents *are* inevitable—but not in *my* organization—not while I'm controlling risk and not when I have the privilege of leading people in dangerous operations. I believe the next accident is always hiding in the unknown, waiting until I least expect it, and suddenly it will attack and try to kill me and my team. I must develop my senses to detect the next potential accident. I must help my team master techniques for operating excellence. If I do this correctly, we can prevent accidents. Even so, we will always need to look over our shoulders for the next accident that is inevitably trying to kill us.

My self-appointed mission is to help operators demonstrate that High Reliability Organization theory is correct—accidents can be prevented. Many managers in various organizations will continue to demonstrate that Normal Accident Theory is correct by succumbing to accidents. This occurs every day in our dangerous world, but it doesn't have to. Humans have a distinct advantage over other animals in the kingdom. We can read accident reports. Sadly though, from reading these reports, humans don't always learn how to prevent the next accident.

Other than in this preface, Normal Accident Theory and High Reliability Organization theory are not part of this book. You can find many good books that describe these theories, and I couldn't improve upon the discussion in any way.

What I have written about is specifically what I have observed for almost forty years in organizations that have been dealing with risk in hazardous operations. Some of these organizations have been highly successful and have accomplished spectacular feats with superb engineering design and excellence in operations. Simultaneously, these same organizations have suffered many accidents along the way, including some large-scale catastrophes. In these organizations, I have seen what has worked and what has not. I have learned from many leaders, the good ones and the not-so-good ones.

Early in my career, I didn't want to be a leader. It seemed too easy to be a bad one. It seemed extraordinarily difficult to become a good leader, to do everything right, to control all the uncertainties, and influence people properly to accomplish missions;

and then, just when I thought things were going well, I could be blindsided by a disaster that I didn't seem to have any control to prevent.

As I continued to observe and think about what methods were successful with certain leaders, I began to notice that the operating leaders, who were leading people in hazardous environments, did things differently than other leaders. I saw a trend in the way they dealt with the hazards and the way they worked with their people. Certainly, every leader has a different style that reflects individual character and personality traits. Playing to his or her strengths, each leader influences people in a unique way.

With all their differences though, the good operating leaders had these similarities: they always demonstrated the highest commitment to the mission, and they cared deeply for the people contributing to the mission. They were acutely attuned to the hazards in the operating environment, they learned quickly how to control the risk, and they were just as quick to share their newly acquired knowledge across to other teams in the organization. Their commitment to the mission, their care for the people contributing to the mission, and their ability to learn and share quickly made these leaders stand out as the best among all leaders. People wanted to follow these operating leaders, especially in dangerous situations.

At each step in my career, I doubted I was ready to be a leader. My boss and mentor at NASA, Mr. George W. S. Abbey, thought otherwise. For twenty years, he put me in leadership situations before I felt I was ready. The aviator's incentive, *learn or die*, seemed easy compared with the leader's incentive, *learn or fail*, with the missions and the lives of other people at stake.

Throughout my careers, in the US Navy, NASA, and the oil and gas industry, I don't recall receiving any specific leadership training in a classroom setting. There may have been some formal training, but, if there was, apparently none of it registered with me as knowledge or skills I needed as a leader. What I did receive was a vast amount of informal, and sometimes unintentional, mentoring. And I learned through observation. As astronaut Steve Hawley has noted astutely, every leader can serve as an example—though, not necessarily as a good example.

I learned much about leadership from leaders I considered unfit for the role. From them, I learned what not to do. Early in my career, I began to realize that some senior officers didn't seem to recognize the poor leadership behaviors of some middle-level officers. Conversely, the beneficiaries of leadership skills—the followers—always seemed to know specifically who the good and bad leaders were. In dangerous endeavors, the followers usually form strong, consistent, and accurate opinions about their leaders.

Mostly, I learned how to be a leader in the field, in trial-by-fire situations. As a twenty-six-year-old Lieutenant Junior Grade in the US Navy, one of my first leadership roles was to lead a group of enlisted sailors who prepared the airplanes in our light attack squadron on an aircraft carrier. At that time, many of the sailors were high school dropouts. One had a previous conviction as a grave robber before he enlisted. He was caught during a scheduled inspection of the barracks cutting up hashish on his footlocker. My job was to be a father figure to him and the other sailors and lead them in hazardous environments on the high seas—and make sure they didn't kill any of our pilots.

Twenty years later, I was assigned to lead a group of 150, type A, overachieving, number-one-in-their-class rocket pilots, engineers, scientists, and doctors, designated as America's astronauts. One of my initial assignments in this role was to restore a culture of operating excellence in the Astronaut Office. In accomplishing this, I first described the *Principles of Operations* for spaceflight crews to codify what had previously been followed but had never been written down. From these, we developed *Techniques for Operating Excellence* to help flight crews execute successful missions and stay alive in the dangerous and unforgiving environment of space. These techniques, which became part of the collective values in our spaceflight culture but were never captured in writing until this publication, are the main subject of this book.

How This Book Is Organized

Controlling Risk—In A Dangerous World contains a large collection of examples, with stories and pictures. Humans learn and remember through stories. The brain is wired to remember and recall interesting, relevant, and descriptive narratives much more easily than dry facts, numbers, and rules. The power of storytelling is well known to military pilots who spend hours in ready-rooms embellishing their tales of valor while waiting for the weather to clear.

Example: Saved by a Story

Here is an example about a story that likely saved my life. In 1979, I was a young and confident naval aviator, returning home from my first deployment with enough experiences to fill several lifetimes. I launched from the aircraft carrier and should have been landing on dry land for the first time in seven months, but the runway ignored its orders and hosted a thunderstorm just prior to my arrival at the welcoming ceremony. After executing an OK-3 Navy landing, demonstrating my superior skill, I made a bad decision on the wet runway and suddenly lost control of my $3 million, single-seat, light attack, A-7 Corsair aircraft.

As I was skidding sideways, headed for an embarrassing death, a colorful tale I heard six months earlier during our deployment flashed into my mind. In an instant, that story helped me save my airplane.

This was the setup for my impending accident. Our squadron had just completed a seven-month cruise aboard the USS *John F Kennedy* in the Mediterranean Sea. As we approached the continental US, we were ferrying our A-7s from the ship to our home base at Naval Air Station Cecil Field in Jacksonville, Florida. During flight operations at sea, the tires on our aircraft were routinely filled with much higher "carrier pressure" to withstand excessive forces during the arrested landings, or controlled crashes. For the fly-off operations, the downside of having higher carrier pressure in the tires was smaller contact area where the rubber meets the runway, resulting in reduced friction and less steering control during landing and rollout on shore.

After launching from the JFK, I joined the maintenance officer's wing for a formation flight to Cecil Field, directly through the heart of the worst thunderstorm I had experienced. Naval aircraft are built to take it. The bigger challenge was yet to come.

On arrival, we were notified the runway was wet. Normal prudence, when landing with carrier pressure on a wet runway, dictated that we should have dropped our tailhooks to take arrested landings. But, the arresting gear system at the field is much less efficient than the shipboard system. The requirement to reset the system after each arrestment would have delayed subsequent landings of the other low-fuel jets coming from the JFK. So my flight leader briefed a new plan to me on the radio. He would land first, and if he experienced difficulty in controlling his airplane on the wet runway, he would notify me, the rookie, to lower my tailhook for a safer arrested landing. If he experienced no problems, I would leave my hook retracted to allow subsequent planes to land expeditiously.

On my final approach, I watched my flight leader land, rollout, and taxi clear with no apparent difficulty. He later admitted to me that he did experience some slipping and sliding but decided not to tell me, thinking that the steering task wasn't too difficult. (Thanks, sir.) I landed on the 8,000-foot runway and had no problems for the first 6,500 feet. Just after I passed the long-field arresting cable, which represented my last opportunity to drop my hook, the automatic antiskid system in my wheels began to shunt hydraulic pressure away from my brakes to prevent lockup as the tires were beginning to hydroplane over the wet surface.

With my current speed and distance remaining and no braking available, I easily calculated I was about to depart the runway without slowing down. Betting that the antiskid system was failing, I decided to try my luck and deselect the system. Of course, I lost my gamble with the laws of physics on slick runways. I should have known that was coming. The antiskid system had been working exactly as designed, doing what it was supposed to do, keeping my steering in control. As quickly as I turned the system off, one tire caught friction while the other continued to slide. My airplane immediately turned ninety degrees to the left, and I was skidding sideways while continuing to track straight down the runway at forty knots, with my right wing pointed forward.

I found myself stable yet out of control, looking over my right shoulder at the end of the runway approaching quickly, without slowing down. After a few seconds, I realized my main wheel was headed directly for an arresting gear stanchion at the runway threshold and mud beyond that. If the wheel dug in, my sideward momentum would flip the airplane on its back. A hilariously good story for other pilots in the ready-room and at my wake.

Instantly, another story I heard in the wardroom on the ship six months earlier flashed through my mind. An F-14 Tomcat pilot was telling us about a dumb action (his words) he took during his landing on a wet runway with carrier pressure in his tires. He blew a tire, lost control, spun around, and somehow ended up traveling straight along the centerline of the runway but backward. Without thinking, he reactively jammed both brake pedals to the floorboard. Bad idea. The rearward momentum of the center of gravity popped the nose of his aircraft up until the tail feathers of his exhaust pipe scraped along the runway. His other main tire blew. Both brakes seized, and the locked wheels ground themselves down to square nubs, as his airplane came to a stop in a spray of sparks.

His crippled plane had to be craned off the runway because they couldn't tow it on square wheels. Through all the laughter in the wardroom, he admitted, "Since I was going backward, rather than stepping on the brakes, all I had to do was go zone-5 afterburner on both engines." At the time, I joined the other pilots and laughed just as loudly at his incompetence. On the inside, though, I silently concluded I never would have thought of that now-obvious solution. What a great story.

Back to my impending death. Without forming any words or taking any time, my brain recalled the relevant part of the Tomcat driver's story, and the automatic processing in my mind quickly invented a solution. I waited until I was approaching the final taxiway in my sideways skid. As the off-ramp reached the two o'clock position relative to my nose, I applied full power to the engine. The big, lazy turbofan spooled up with its usual delay, and by the time the taxiway was at my one o'clock position, sufficient thrust was beginning to build to push my airplane toward the taxiway. The plane exited the runway straight onto the centerline of the last taxiway before disaster. I retarded the throttle quickly, and the A-7 gently skidded to a stop, as if I had planned my graceful slide all along.

After I got my heart rate below 100 bpm, I said a silent prayer thanking the F-14 pilot for telling his tale of misfortune. His story saved me.

That's why I have decided to fill this book with examples and stories. I hope they help you control risk. As you travel through life in your dangerous world, be observant. Develop your stories based on your experiences, both good and bad. Share your stories with others. Help them save lives.

Terms Used

I use some terms extensively in this book:

- *Organizations* or *companies* are groups of people collectively working to achieve a mission. In a dangerous business, the organization or company is a large, complex *sociotechnical system* the people use to: (1) conduct activities intended to deliver results in service of the mission and (2) manage and control risk to prevent accidents.
- *Leadership* is not a person or the team at the top of an organization. As I use the term, leadership is the skill leaders use to influence people to take actions and make decisions.
- *Leaders* are the people who are designated in their organizations to oversee one or more people. Leaders use their skill of leadership to motivate and inspire people to accomplish more in service of the mission. Leaders include *influence leaders* who have not been officially designated, but, by virtue of their demonstrated actions in influencing others, are sometimes called influence leaders.
- *Operators* are people who control or operate systems; in hazardous environments, these are the people who are confronting hazards. Examples of operators are pilots, crewmembers, front-line workers, doctors, nurses, construction workers, drillers, roughnecks, sailors, soldiers, Marines, and many others.

I write about three levels of leaders in the organization, using designations as follows:

- Top level—executives, senior leaders, and senior managers
- Middle level—managers
- Lower level—front-line or first-level leaders, team leaders, and supervisors

Chapter Summaries

Chapter 1 is about how organizations manage risk. Some of the content covers how I think managers *should*, and sometimes *do*, collectively manage risk in organizations to prevent accidents and help their workforce improve performance. Some content in the chapter is based on my personal observations and opinions of what the managers did and how they were attempting to manage risk before they failed to prevent accidents or simply caused poor performance in their organization.

As the first chapter shows how managers manage risk, chapter 2 illustrates how operators control risk. I have written chapter 2 with the perspective of the front-line operator who is facing hazards every day. Operators think about risk differently than managers. Managers manage risk. Operators control risk. Managers can change the design of equipment or a system to reduce the hazards. They can use various probabilistic analyses to calculate the risk and decide if the risk is below an acceptable level and what actions are required to monitor the risk. Operators don't have the luxury of redesigning or changing the system, and the probabilities of being injured are irrelevant. Operators must use the system given and must face the dangers every day and try not to be injured, regardless of the probabilities calculated by a manager in an office. Every operator is in the last line of defense protecting the organization from disaster.

Risk attitude may be the single most important characteristic needed for an operator to control risk successfully. Notionally, risk attitude can be thought of as the personal ratio of risk perception to risk propensity. Operators with the best risk attitude will be those who have a great ability to sense risk and a low desire to accept that risk. The risk attitude harbored by an individual is difficult to *quantify* numerically. With experience, though, the *quality* of risk attitude is easy to judge.

Consider this:

I believe I can predict which operators will not die when operating with high risk. For example, among the elite big-wave surfers, I have confidence Laird Hamilton will not die while surfing. I don't know the man and have never met him, but I have read his statements and a description of his risk philosophy in *Sports Illustrated* magazine.[3] Some surfers want to master the ocean and ride the biggest wave. Hamilton believes it is the ocean that allows the surfer to ride the wave, or not. To survive, the surfer has to be smart enough to know when the ocean does not want to be ridden. Hamilton does not attempt to master the ocean. He accepts the privilege of riding when the conditions allow the waves to be ridden. Additionally, Hamilton and his team spend much more time practicing rescue techniques than other surfers. They demonstrate the right attitude.

On the other side of the spectrum of risk attitude, I found Maurice and Katia Krafft, who were volcanologists, or volcano chasers. While watching a documentary about them on television[4], I listened to their quotes. Before the broadcast ended, I

---

3   Jeff MacGregor, *Sports Illustrated*, February 25, 2000.
4   *Maurice & Katia Krafft: Lives on Fire,* television documentary; American Public Television, distributor; Europe Images International, producer; rights, contract period April 1, 2011–March 31, 2013, expired (not in rights).

concluded they would not survive in their work. Read the quotes I transcribed and decide for yourself:

Maurice: "For me, active volcanoes, especially ones I know very well, they are like friends."

Maurice: "We have seen bigger and bigger eruptions; the more enormity, the better it is for me."

Katia: "For me, when I'm working around the volcanoes, I forget everything and there is no more danger for me."

Maurice: "I'm never afraid because I have seen so much eruptions in twenty-three years, zat, um, even if I die tomorrow, I don't care."

Near the end of the program, I learned that after ignoring warnings they traveled to within two miles of Japan's Mount Unzen in June 1991 and were killed in an eruption.

Sometimes, not much separates the survivors and the nonsurvivors. Risk attitude, though, is one delineator. As a final example, see the following table, which compares the risk attitudes of two kinds of risk takers: BASE jumpers (who leap from Buildings, Antennae, Spans, and Earth—for fun) and astronauts (who also leap from the earth but in the other direction and for a different reason).

	BASE jumpers	**Astronauts**
Risk-taking Goal	Get as close to death as possible and still survive	Stay as far away from death as possible
Motivation	Adrenaline rush (Risk-driven)	Accomplish the immediate mission and many more missions in the future (Mission-driven)
Amount of Training	Less preparation *"That's enough; let's go."*	More preparation *"We've thought of everything; we're ready."*
Philosophy of Death	Accept death, if it comes *"The only time I'll stop is if I die." "If I am going to die – what a great way to have lived."*	Accept death, if there is no way out, but will fight to stay alive *"If I'm going to die, I will spend my last seconds trying to save my crew, our vehicle, and the mission."*

Table 1. Risk Attitude of Two Kinds of Risk Takers

Chapter 3 details some of the work we conducted in the Astronaut Office at NASA to develop and codify the *Principles of Operations* for spaceflight in an effort to prevent another fatal accident. Mr. George Abbey, one of the best operating leaders in hazardous endeavors, is the father of the International Space Station, flying safely and productively as I write this, fifteen years after its launch. In 1998, his intuition, based on his extensive experience and his ability to detect minor signals in how astronauts were operating, indicated the Astronaut Office was headed for trouble. Even though we were staffed with some of the best aviators on and off the planet, he felt we needed to improve the way we were operating. As it is with all good safety decisions, no one can prove his insight was correct, that we were headed for an accident. We took positive actions to improve the way we operated on the ground, in the air, and in space. We did not have a fatal accident under his leadership.

Chapters 4 and 5 are the main sections of this book. These are where I describe thirty *Techniques for Operating Excellence*. My intent is for you to use these techniques to prevent your next accident, save lives, improve performance, and achieve more than you thought possible. I believe these techniques are applicable in all endeavors involving risk. Modify the techniques as necessary for your application, develop your own stories to share with others, and improve the techniques as you learn what helps you and your team to operate successfully. The techniques I present in chapters 4 and 5 were based on the Principles of Operations for spaceflight crews, listed in chapter 3. Five decades before these thirty techniques were captured in writing, the original Mercury 7 astronauts, who were selected from the best test pilots in military aviation, used similar techniques to achieve operating excellence in space.

Four appendixes are included. They contain individual techniques that managers can use to influence and inspire operating excellence in specific situations. The subjects are:

A. Seven Leadership Principles
B. Operating Leadership Behaviors
C. Creating Commitment and Accountability
D. Policy Note—Astronaut Office Conduct and Performance

Finally, as I close the preface, I leave you with this thought. Someday I will die. I intend the cause of my death will be old age, not some tragic accident. (Writing this paragraph reminds me of the Will Rogers quote: "When I die, I want to die like my grandfather, who died peacefully in his sleep. Not screaming like all the passengers in his car.") Though the personal Techniques for Operating Excellence can't help me live forever, the techniques are intended to help me accomplish much more in this dangerous world, while preventing the next potential accident until I reach my expiration date as a very old controller of risk. If I die prematurely, in some fiery, preventable accident, it doesn't mean the techniques were wrong. It only means I didn't execute them well enough to control the risk. And it will be left to you, to master the techniques, improve them a little, live a bit longer, and pass along your wiser, risk-controlling genes to the next generation.

Chapter 1

MANAGING RISK
An Organizational Responsibility

Creating Success in Dangerous Operations

Exploring space for knowledge, or Earth for oil, involves inherent hazards. Incredibly complicated operations must be conducted in volatile environments. Organizations can quickly create dead explorers by failing to prevent every potential accident.

Humans, with their innate strengths in learning and adaptability, are well suited to manage these dynamic and dangerous operations above, below, and on the surface of the earth. To tap into the collective wisdom of humans working together, managers are grouped into organizations that use complicated sociotechnical systems to control the volatile operations while continually trying to improve production and prevent accidents.

Modern-day missions are so complicated and dangerous, and are being conducted in such dynamic and obscure conditions, that no single person has sufficient capability to manage the whole operation and control the risk. Decision making is distributed among managers and personnel in the organization. Control of operations and risk is accomplished with a wide and deep sociotechnical system. The problem, though, is that this system is more complex than the operations it is intended to control.

As the name implies, two parts work together in the massive sociotechnical system controlling operations and creating productivity. There is the *social* side, with engineers, managers, and operators, who must develop relationships and communicate well with one another working together as a social group to make the best decisions. In complicated, hazardous operations, a committee does not make decisions. Single managers have the responsibility to make specific decisions in any given operation. But those decisions must be informed by relevant inputs from knowledgeable people who are distributed in the complex sociotechnical system. Subsequent decisions will become less effective over time if the social relationships begin to break down on the human side of the sociotechnical system.

And there is the *technical* side feeding technical information to the humans in the system who are making decisions and taking actions as they conduct operations. The technical side can be thought of as everything in the organization not human. It is the large, complex collection of rules, policies, and procedures; it is all the equipment, hardware, software, firmware, control systems, instrumentation, and sensors; it is the various mechanical processes for operations and procedural processes for risk management and control of hazards; it is the training programs for workers and assurance processes for managers. The list of items on the technical side is longer than I have documented here.

Before and after accidents, managers in many organizations focus their attention on trying to improve the technical side of the system to control risk and improve productivity. Far too little effort is spent on improving the social human side, which has the power to create an exceptional organization and support great human achievement. When too little attention is devoted to this human side, the resulting ineffective relationships, poor communication, and bad decision-making have the ability to create tragic outcomes and destroy an organization.

This is a book devoted mostly to the social human side of the complex sociotechnical systems used by organizations involved in dangerous endeavors. In such complex systems, reliable performance calls for decision-making with good judgment under uncertain, complex, and ambiguous conditions with time pressure. How successfully and sustainably over the long term humans can explore and accomplish missions will depend upon the ability of their leaders to learn, adapt, and make effective decisions based on superior judgment, vast experience, and high values.

As humans individually and collectively learn and become skilled, tasks related to the skill are quickly relegated to automatic mental processing. There is a short span of time between learning a skill and no longer needing to think about the specific tasks required to perform that skill. The ability to learn a skill serves as a great evolutionary and organizational advantage. The human brain can process higher-order executive functions when it no longer is constrained to controlling lower-level motor functions and thought processes required in performance of the skill. This ability to process more inputs and greater information represents the incredible power of the human mind. Individual managers get smarter, and the organization collectively gets better.

But the decreased cognitive attention to the lower-level functions after becoming skilled has an occasional downside in humans—and in the collective organization. In the individual, when the brain is engaged with higher functions, the decreased cognitive attention to the details of the learned task can result in overlooking or missing some of the lower-level steps in a complex operation. Even though the operator's brain is engaged with important higher-order functions, the failure to pay attention to the details of a hazardous operation can result in disaster. Skilled people get complacent. Then they die.

The human strengths of learning, adaptability, and the ability to process more information after becoming skilled are the very same strengths that organizations use to create successful performance in complex operations. Through success over time, the managers and operators become collectively skilled at controlling the hazardous operations and achieving high-quality results.

But an organization can, and often does, exhibit a kind of collective complacency. With accumulating success over time, the skilled managers and operators, who think they have learned how to control the system, stop learning and adapting. Eventually, this collective complacency in the organization causes the sociotechnical system, which is intended to control increasing risk in dynamic and dangerous environments, to become less dynamic and adaptable. The complacent system exposes the organization to a particular vulnerability with slowly increasing risk—an almost imperceptible *drift toward the next accident.*

How Accidents Emerge in Organizations

In Sidney Dekker's perceptive book, *Drift into Failure*, he describes what really happens in organizations before major accidents. Based on my observations and experiences in organizations, he correctly describes what causes organizational failure. Here's what he writes and how it applies to controlling risk.

> "Incidents do not precede accidents. Normal work does. . . . [Accidents] cannot be predicted on the basis of the constituent parts [of complex systems]; rather, they are one emergent result of the constituent components doing their normal work."

> "Organizational decisions [before accidents]. . . seemed like perfectly good or reasonable proposals at the time. . . . [Decisions are] seldom big, risky events or order-of-magnitude steps. Rather, there is a succession of weak signals and decisions, a long and steady progression of small, decremental

steps of accepting them that unwittingly take an organization toward disaster. Each step away from the original norm that meets with empirical success (and with no obvious sacrifice of safety) is used as the next basis from which to depart just that little bit more again. It is this decrementalism that makes distinguishing the abnormal from the normal so difficult."[5]

Dekker also writes, "accidents emerge from these relationships [in the system], not from some broken parts that lie in between." He includes a quote from Rasmussen and Svedung's *Proactive Risk Management in a Dynamic Society*: "Accidents are the effect of a systematic migration of organizational behavior under the influence of pressure toward cost-effectiveness in an aggressive, competitive environment."[6]

For long-term viability of a company, a profitability motive is always desired. But success under a continuous emphasis on profitability requires actively controlling the sociotechnical system and detecting decremental changes in the relationships between parts of the system to prevent drift toward the next accident.

Accidents happen after systems experience slow degradation. Event-based investigating only shows *what* happened before the previous accident. Predicting and preventing the next accident requires an understanding of *how* and *why* the system is degrading. This is the realm of systems-based investigating.

Investigating Accidents

After an accident, organizations typically conduct investigations to determine the causal events, including the immediate, or proximal, cause that led to the accident. Many of these event-based investigations consider the determination of a root cause to be an important conclusion in understanding what went wrong. But Dekker suggests that trying to identify a root cause is pointless. In his book, *The Field Guide to Understanding Human Error*, Dekker writes, "What you call 'root cause' is simply the place where you stop looking any further."[7]

After causes are identified through event-based investigating, corrective actions are developed to help the organization prevent future accidents. Usually, these corrective actions include constraints in the form of new rules and procedures for operators to follow. These rules-based procedures may be successful, for a while, in preventing similar potential accidents from occurring under similar conditions from known causes.

---

5   Sidney Dekker, *Drift into Failure* (Burlington, VT: Ashgate Publishing Company, 2011), 48, 49, 63, 158, 188.

6   Quoted on page 1 in ibid., J. Rasmussen, and I. Svedung, *Proactive Risk Management in a Dynamic Society* (Karlstad, Sweden: Swedish Rescue Services Agency, 2000), 14.

7   Sidney Dekker, *The Field Guide to Understanding Human Error* (Burlington, VT: Ashgate Publishing Company, 2006).

When managers rely on event-based investigating, which identifies only the causes of past known accidents, the organization will have difficulty in preventing future unknown accidents. Believing accidents are caused by a limited number of specific events is a fallacy. Hindsight bias leads some managers to think they can use a straightforward retrospective analysis to identify lines of causality that clearly point directly to the accident.[8] Then they wonder how the victims missed those obvious signals.

Accidents are rarely so simple. Hardware does not simply "just break." People do not make mistakes based on simple individual previous events. The next accident won't be like the previous accident. After investigations, more rules are promulgated, which can overload operators and constrain good judgment. In the ever-changing operational situations, the new rules become out-of-date, confusing, ineffective, erroneous, and ignored.

If the next potential accident were similar to a previous accident, the next accident and its bad consequences could be prevented easily. At the same time, though, there will be hundreds of different unknown potential accidents targeting the organization, giving thousands of hard-to-sense signals from unpredictable directions, all vying for attention yet getting lost in the cacophony of everyday operations and more rules. Preventing every single one of these unknown potential accidents, which are coming every day, is the foundation of operating excellence. Avoiding tragedy, while performing noble missions at peak effectiveness, is the holy grail of organizations trying to operate in hazardous environments.

So, what should an organization do after an accident to investigate effectively, learn lessons, and implement the proper corrective actions to prevent future accidents?

Managers should commission an event-based analysis initially to determine what occurred. Investigators determine retrospectively what decisions the managers made and what actions the operators took before the accident. The conclusions are linear and causal. After understanding *what* happened through the event-based analysis, the organization should conduct a valuable *systems-based analysis* using systems thinking to determine *why* those events occurred.

A systems-based analysis yields better learning. Investigators can determine why the decisions made and actions taken seemed appropriate to the managers and operators who were under the influence of the sociotechnical system. Corrective actions can be developed to improve the sociotechnical system so managers and operators in the future will not be influenced in the same way they were in the past before the previous accidents. Corrective actions generated from a systems-based analysis have a greater chance of helping the organization prevent future accidents.

---

8   From a personal conversation with Jim Reason, Professor Emeritus, University of Manchester, UK, on October 27, 2004, in Cleveland, Ohio, USA, at a Risk Conference hosted by NASA.

By understanding the systemic conditions in place that led to previous accidents, leaders and personnel can tune their senses to detect similar conditions and either improve the conditions or control the system influences to prevent future accidents. Systemic issues or adverse conditions, in one or more of the following areas, have proceeded past accidents and will precede future accidents unless corrected or improved:

- Complexities of the sociotechnical system and its interfaces
- Organizational culture and climate
- Managerial influences or pressures that affect decision making
- Challenging operating conditions

If these conditions are not understood well, the organization likely will not be able to prevent the next accident. The vigilance required to sense problems in complex organizational conditions is demanding and makes the responsibilities of a leader quite challenging. For me, though, developing the skills to identify those adverse conditions and successfully predict and prevent accidents, while leading people in hazardous environments, is rewarding.

Drivers, Derailers, and Catalyzers— The Behaviors That Influence Performance

After spending years in various organizations that had experienced small incidents, minor accidents, and major disasters, I began to think about how the top-level executives, middle-level managers, and front-line supervisors influenced operations and organizational performance with their words, decisions, and actions. I hold a fundamental assumption that people are honorable at all levels in organizations, and no one is attempting to make bad decisions or take actions intended to cause damage. Rather, people generally try to make good decisions and take appropriate actions consistent with their understanding of their role in the organizations.

So, where and how does it sometimes go horribly wrong? How does an organization slowly drift from high performance to catastrophe? Does the degradation occur at the executive level, the managerial level, or the front-line level in the organization, or at all three levels simultaneously? By the way, in the years I have spent in organizations before and after major and minor accidents, I have heard members in each of these levels in the organization blame the personnel in another level for causing the poor performance and accidents.

Most of the time, executives, managers, and front-line supervisors say and do the right things. At each of these levels, the leaders have certain roles, skills, expectations, goals, concerns, constraints, and limitations unique to their level. These level-specific influences drive what the leader says and does. Executives behave like executives because they have executive-level *drivers* based on what they are expected to accomplish as executives. Their behaviors are predictable. This is similar for middle-level managers and front-line supervisors. Usually, these drivers push the leaders to say and do the right things to support the organization's goals.

But humans are not perfect. As the organization matures, organizational and personal stressors begin to degrade the performance of the leaders at all three levels simultaneously. The stress can come from various outside and inside influences, such as market changes, competition, loss of talent, multiple accidents, reorganization, production incentives from bosses, or any number of potentially negative influences. The organizational stress causes each leader to feel individual pressure to perform better.

Two things happen that begin to derail the organizational train, as the leaders at each level respond to the stressful influences.

First, collectively, the leaders in the organization begin to define the organizational mission in terms of short-term performance, rather than long-term performance and enhancing the viability or health of the organization. Short-term performance gains make the organization feel successful, or at least, feel better about its results. With the shortcuts, though, long-term performance and results suffer. For example, preventative maintenance may be deferred until next quarter, or next year, to make the current performance numbers look better. The longer this practice of deferral continues, the harder it is to dig out of the organizational hole of decreasing performance and poor long-term results.

Second, in an individual response to the stress, the leaders in each organizational level are motivated personally to do their jobs better and better. Without guidance from the executives specifying that good performance be defined by how well individual leaders work together in support of the overall mission, individual leaders begin to be motivated by demonstrating better self-performance. As a motivational influence, personal goals begin to outweigh the organizational mission. Teamwork suffers. Organizational performance suffers. Missions are harder to achieve because leaders are trying to look good to their bosses rather than making effective decisions for the good of the team and the mission.

The organization responds to stress, first, by redefining the mission in terms of short-term performance, and second, by allowing personal goals to outweigh the organizational mission. From these two responses, *derailing* behaviors begin to show up from leaders at all three levels in the organization, simultaneously. These derailers occur simply because the leaders are trying to do their jobs well, but they succumb to the allure of being recognized for individual and short-term accomplishments rather than continuing to make thankless contributions to the long-term mission.

The best organizations are highly introspective and self-reflective. They analyze their performance and learn quickly. The leaders in these organizations understand the normal *drivers* that create good performance. They know self-centered *derailers* executed for short-term gains will naturally follow and degrade the long-term performance of the organization. Ultimately, the good leaders develop ways to mitigate the derailing influences to *catalyze* a return to better, sustainable performance. These catalyzing actions will eliminate accidents, improve long-term performance, and allow the organization to accomplish its goals and contribute to society for many years.

The following three tables show what influences the decision making of the leaders in the three different levels of the organization. The columns in each table represent three different stages of organizational maturity, whether the leaders respond more to drivers, derailers, or catalyzers.

The first column in each table shows the drivers, or what influences the leaders as they make effective decisions in pursuit of the mission. When the leaders are operating in the state described in this first column, the personal goals of the leaders in each level are aligned with the organization's mission. Each leader is doing the right thing to help the organization accomplish its mission.

The second column shows what can derail or adversely influence decisions when the organization is under stress. The behaviors shown in this column result from the two general changes that occur when the organization becomes stressed:

1. Redefining success in terms of short-term performance, and
2. Allowing personal goals to outweigh the organization's mission.

These resulting behaviors, if unchecked by behaviors in the third column, will result in a slow, unconscious, organization-wide, decremental drift toward the next accident.

The third column suggests how catalyzers can mitigate the derailing influences and accelerate performance to new levels of operating excellence. In high-hazard organizations, leaders use these catalyzers to make effective decisions, improve performance, and manage risk.

Leaders (Level in the Organization)	Drivers Desired behaviors expected to achieve good performance; Personal goals aligned with organization's mission	Derailers Adverse behaviors resulting from organizational stress 1. Success = Short-term Performance; 2. Personal Goals > Mission	Catalyzers Necessary behaviors to mitigate derailers, prevent accidents, and improve performance
Executives (Top)	• Encourage profitability • Enhance long-term value • Incentivize accomplishments • Strive to increase outputs, results • Avoid major accidents	• Excessive profitability motive • Satisfy investors (short-term) • Drive for ultimate optimization (faster, better, cheaper) • Focus on results, ignore quality of process • Not worrying about minor accidents • No longer searching for vulnerabilities • Give ambiguous signals • "Work harder; just get it done."	• Temper profitability motive • Incentivize improvements in long-term value • Allow some slack in optimization • Attend to inputs, quality of process, ways of working; "Work smarter"[9] • Invest in safety and improvements • Provide decision-making guidance

Table 2. Drivers, Derailers, and Catalyzers for Executive Level (the behaviors that influence decision-making)

9   From conversations with Nelson P. Repenning, Professor of Management Science and Organization Studies, Massachusetts Institute of Technology (Boston, MA, 2012)

Leaders	Drivers	Derailers	Catalyzers
(Level in the Organization)	Desired behaviors expected to achieve good performance; Personal goals aligned with organization's mission	Adverse behaviors resulting from organizational stress 1. Success = Short-term Performance; 2. Personal Goals > Mission	Necessary behaviors to mitigate derailers, prevent accidents, and improve performance
Managers (Middle)	• Improve performance *and* manage risk • Accommodate drivers from executives and front-line leaders • Make decisions for benefit of company • Earn a bonus and approval from boss	• Improve short-term performance • Reduce risk perception (of risk across teams, and accumulation of future risk) • Increase risk propensity (tolerance to accept risk) • Make decisions without front-line inputs • Manage upward relationships to satisfy bosses • Encourage competition • Punish mistakes and low performance • Make decisions to benefit only my team or me (defer preventative maintenance, etc.) • Misread signals and messages from the top • Fail to create accountability before the incident	• Make conservative decisions for good of the mission, using: 1. Values, judgment, experience, leadership behaviors 2. Technical rationale, proper processes, criteria, responsibilities • Manage downward relationships; support success of teams • Reward cooperation and learning • Increase risk perception • Decrease risk propensity • Develop and provide good procedures • Conduct after-action reviews • Reward good practices and following processes of operating excellence

Table 3. Drivers, Derailers, and Catalyzers for Middle-level Managers (the behaviors that influence decision-making)

Leaders (Level in the Organization)	Drivers Desired behaviors expected to achieve good performance; Personal goals aligned with organization's mission	Derailers Adverse behaviors resulting from organizational stress 1. Success = Short-term Performance; 2. Personal Goals > Mission	Catalyzers Necessary behaviors to mitigate derailers, prevent accidents, and improve performance
Front-line Leaders (Lower)	• Focus on job/task success • Identify hazards, control risk, avoid injury • Enhance cooperation • Attend to inputs, process • Be present, stay in the moment, be mindful	• Allow practical drift[10] ("We don't use the procedures; we follow a better way.") • Encourage "Get it done" mentality • Fail to recognize and prevent complacency • Violate safety rules or accepted processes to accomplish jobs • Take more risk	• Monitor and control practical drift • Use good procedures; employ best practices • Combat complacency • Follow principles and techniques for operating excellence • Follow safety rules

Table 4. Drivers, Derailers, and Catalyzers for Front-line Leaders (the behaviors that influence decision-making)

---

10   "Practical drift—the slow, steady uncoupling of practice from written procedure;" a term originated by Scott A. Snook, *Friendly Fire: The Accidental Shootdown of U.S. Black Hawks over Northern Iraq* (Princeton, NJ: Princeton University Press, 2000).

Ten Adverse Conditions in Organizations
That Failed to Prevent the Next Accident

Based on my observations, discussions, and assessments, I created a list of ten common conditions that existed in various organizations before they experienced major disasters or minor accidents. In the sociotechnical system used to control risk, improve performance, and accomplish goals, five of these ten adverse conditions existed on the technical side, and five were on the social side. By turning hindsight into foresight, good leaders can use the presence and severity of these adverse conditions as indicators to determine if their organization is drifting toward a disaster.

Technical / Systems / Managerial Side

1. Emphasized Organizational Results Rather Than the Quality of Individual Activities
Individual people in an organization don't create *results*; they conduct *activities*. Even groups of people performing multiple activities don't create results. The multiple activities of groups are still anchored in the present. All the in-the-moment activities—integrated over time—create results. High-quality activities, when added together, result in high-quality results. In a dangerous business, a single poor-quality activity can result in disaster.

Results *are* important, but the quality of activities creates the quality of results. Executives and managers should talk to their front-line supervisors and workers about results. But the workers want to know how their individual activities are contributing to those results and, more importantly, how to modify their activities if the results are not meeting expectations. In organizations drifting toward disaster, the executives and senior-level managers emphasized results and didn't mention, or pay attention to, the quality of activities.

Without guidance, motivation, or incentive to conduct high-quality activities, the middle-level managers, front-line supervisors, and workers began to compromise the quality of activities to achieve results more expeditiously to satisfy the executives and senior-level managers. Common phrases heard around the workplace were "get it done" and "work harder." Workers talked about "schedule pressure" and "cost pressure" levied from above. Speeches from bosses often contained analogies from sports, where competition was emphasized, as in, "Next quarter, we must beat our competitors and win decisively." Discussions about the quality of activities, cooperation, working together, and learning to work smarter were absent.

2. Stopped Searching for Vulnerabilities—Didn't Think a Disaster Would Occur
Managers usually thought their organization and teams were performing well just before the disaster occurred. Managers were unaware of some performance

issues. Worse than their incorrect assessment of performance, the managers had simply stopped searching for any performance issues or weaknesses in the organization.

Accident reports typically listed immediate and contributing causes to an accident, as determined by safety investigators who conducted retrospective analyses of evidence. In reports of major accidents, investigators commonly described specific failures that were not anticipated or identified by the organization before the accident.

I observed the situations were even worse. Not only did the managers fail to identify the *specific* weaknesses or system breakdowns that occurred; senior-level managers had stopped searching for *any* weaknesses or vulnerabilities in components, policies, processes, competence, relationships, communication, or any other part of their systems or people in the organization.

After the *Columbia* tragedy, our *third* accident that resulted in the deaths of spaceflight crewmembers, a respected manager explained to me how it happened by saying candidly, "We just forgot how dangerous spaceflight really is."

3. Didn't Create or Use an Effective Assurance Process

Prior to accidents, managers in organizations did not understand how to create an effective process of assurance, nor did they understand the value of assurance. An effective assurance process was not in place or used.

No single organization can conduct all the operations required to accomplish complicated objectives, such as constructing the International Space Station in orbit. Many organizations and groups must contribute to the mission simultaneously. Making strategic decisions is more challenging for the executives, who must be able to predict future progress toward mission goals with multiple organizations. Unfortunately, no single person is capable of understanding how well each entity is performing in service of the mission. To succeed—and prevent disasters—executives must rely on an effective assurance process to help them predict the performance of each entity.

In an operational organization, providing assurance means a person is giving confidence about future performance to another person, or group, based on observations or assessments of past and current performance. An assurance team observes a sample of operations and evaluates the quality of performance. From their assessment, based on their professional experience, the assurance team leader can give an executive (or governing board) confidence the assessed entity will continue to deliver desired performance in the future—or won't continue. That confidence should be qualified with derived and estimated uncertainties, based on how detailed and complete the observations or inspections were when assessing the quality of past and current performance.

Before major accidents, executives and managers did not understand how well—or how poorly—supporting organizations were performing because they did not create or use an effective assurance process. Without this, the executives had erroneous confidence that organizations would continue to perform well.

4. Allowed Violation of Rules, Policies, and Procedures

After accidents, investigators usually determined that some organizational rules, policies, and procedures were violated before the accident. Often, the workforce reported unofficially that some managers were cognizant of these violations in operations before the accident. In some cases, according to the workforce, managers condoned violations in an effort to entice greater production or faster results. Front-line workers were willing to take excessive risks to satisfy their managers.

5. Some Leaders and Operators Were Not Sufficiently Competent

I observed inadequate and misjudged competence in some executives, managers, supervisors, and operators before major accidents. From inflated self-assessments, their confidence was erroneously high and unjustified. Most leaders were unwilling to admit they didn't have all the answers. Poor decisions were made. Preventing major accidents in complicated operations requires high levels of managerial and technical competence in all personnel.

Deficiencies in knowledge, skills, or attitudes at any level in the organization can result in a failure to prevent accidents. Qualified assessors should have been assigned to test knowledge, assess skills, and evaluate attitudes of all people who were contributing to the hazardous missions.

Each employee should endeavor to:

- Know his or her job better than the supervisors and auditors;
- Know the rules better than the safety inspectors and regulators; and
- Know the hardware, software, and systems better than the engineers or designers.

Each employee should be able to demonstrate skill through practical application of knowledge in real or simulated hazardous environments.

Executives and managers in the operating chain of command should have been responsible for evaluating the knowledge, skills, and attitudes of their people. The executives and managers did not fulfill this responsibility effectively.

Social / Human / Leadership Side

1. Leaders Didn't Fulfill the (Transformational) Purpose of Leadership
Before accidents, most leaders understood the purpose of leadership only in a *transactional* sense. That is, they understood their role was to create a vision, develop a strategy, provide guidance, issue directives, create schedules, procure supplies, conduct training, and handle various other *one-way* transactions. The operating workforce was already creating value and advancing the mission, in addition to facing the hazards every day on the front lines of operations. Likely, the workforce would have continued to do their jobs with or without the leaders' transactional guidance.

It was the rare leader who understood the *transformational* purpose of leadership: *The essence of* transformational *leadership is enabling, motivating, and inspiring a group of people to perform better individually and accomplish more collectively with higher quality, in service of a mission or pursuit of a goal, than they would have without the leader's influence.*

Leaders should inspire their people to create success. Before minor accidents and major disasters, the leaders did not inspire their people sufficiently through transformational leadership.

2. Leaders Didn't Create Accountability (nor Commitment) Before the Incident
The concept of accountability was not well understood. A manifestation of accountability was often executed *after* an incident. The managers held a person or group to account after the incident by applying adverse consequences, such as blame, letters of discipline, reassignment, or termination. Almost never did I see accountability being created *before* an incident. I rarely saw accountability being managed in a productive way.

The root of the word *accountability* is to give an account, or provide justification to someone. A leader should create accountability in subordinate leaders and workers by periodically causing them to answer questions and give an account of past and present performance and future predicted performance.

The two-person (at a minimum) process of exercising accountability should be accomplished on a routine basis in a supportive way. The managing supervisor should create accountability by establishing a schedule to meet and discuss operating performance and delivery of objectives with the person who is being held to account, that is, the subordinate manager, team leader, or operator. The managing supervisor should require the person to provide an accounting regularly and answer questions similar to the following: What did you and your team accomplish last week? How well did your team perform? What did you and your team learn? What issues have you uncovered? What is your schedule and predicted performance in the near future? What obstacles do you foresee? What resources do you need? How can I support you?

When the subordinate manager or operator experiences this kind of interaction she or he feels accountable to the supervising manager. Generally, operating performance and safety improve when a person feels accountable for performance to another person or group, such as a corporate board, the press, or the public. The process of accountability works best if the person responsible is periodically called upon to provide an account to someone.

In a hazardous business, success depends upon a culture of commitment and accountability throughout the leadership chain of command. If the behaviors— and knowledge, skills, and attitudes—of any employee are not aligned with the values of the organization, that employee must be moved immediately to a position in the organization where he or she can provide value. No apologies for the move are necessary. If the senior leader truly cares for the manager or operator who is demonstrating misaligned values, the senior leader should not want to leave that manager or operator in a situation where he or she can fail or damage the mission.

I saw accountability exercised extensively *after* incidents or errors, and exercised rarely during normal work *before* incidents. See appendix C for a description of how a supervising manager can create accountability and commitment with a subordinate manager, supervisor, or operator.

3. Leaders Didn't Sufficiently Listen To, Engage With, or Care For Their People

Large organizations tended to forget people were doing the work, not machines or systems. Managers tended to focus on improving the systems, but the managers should have helped the people who were trying to use the systems.

Often the front-line workers understood the fragile, volatile, and uncertain state of operations in complex and dangerous endeavors. If the managers had engaged in a better way with their people, major accidents might have been predicted and averted. Additionally, the performance improvements the managers were pushing for might have been realized. Performance goes up when leaders tap into the collective energy of people who feel supported while working together.

Before accidents, executives and managers often pushed people well past their individual red-line limits. Executives and managers didn't really listen to their workforce. After the *Columbia* accident, a senior executive defended the excessive pressure he was exerting on the workforce to accelerate the schedule by telling me, "If we back off on the schedule, they will relax." In my opinion, that executive had no idea how hard the workforce was already working as they put their hearts and souls into the mission of launching people into space.

4. Some Leaders and Operators Placed Self-Interests above the Organization
Organizations and groups that performed poorly generally had individuals who displayed little or no sense of duty to the mission. Such people or groups were more concerned with the politics of the organization than supporting the mission or teams. Some people were motivated more by getting promoted, receiving bonuses, and how they appeared to their bosses, rather than executing their responsibilities in a high-quality way. Because of self-interest, some managers and operators were reluctant to admit they didn't have all the answers, which led to misplaced confidence, poor decisions, and damaging actions. Often in organizations that experienced accidents, people and teams didn't possess a sense of duty to the mission. They were not helpful and didn't share valuable information, knowledge, or experience with others in the organization.

5. Leaders and Operators Didn't Possess Error Wisdom[11]
In organizations headed for disaster, managers and operators rarely understood patterns of errors or biases in decision-making in themselves and their teams. They rarely dedicated time and effort to analyze errors and decision-making. Errors were seen as a sign of individual weakness rather than possible indications of systemic deficiencies in the organization. I saw insufficient efforts to help operators develop techniques to reduce the likelihood of errors before they occur, capture errors as they occur, and mitigate the consequences of errors after they occur.

All humans make mistakes. High-performance teams understand this and embrace learning from mistakes. The best organizations involved in hazardous activities have acquired error wisdom and have developed techniques and systems to eliminate or reduce the consequences of errors.

Successful Leadership in Hazardous Environments
The energy industry and the space program have inherent dangers. Companies must operate in hazardous environments to accomplish the goal of delivering products and services to society. Space is an unforgiving environment that will extinguish life in an instant. For long-term sustainability, an organization operating under dangerous conditions must simultaneously *improve performance* and *manage risk*, or it will die.

The challenge for leaders in any organization is to improve production results while staying in the parity zone (illustrated in the notional chart below, from J. Reason) with appropriate amounts of production and protection, or safety. The goal is always to operate within the boundaries of catastrophe and bankruptcy. Human and mechanical systems and interfaces are complex, and the environments

---

11   "Error wisdom" is a term James Reason used (*Beyond the Organizational Accident: The Need for "Error Wisdom" on the Frontline* [Quality Safety Health Care, December 2004]).

in which they operate are dynamic. The boundaries of catastrophe and bankruptcy are not static and may always be impossible for humans to define or discern at any moment.

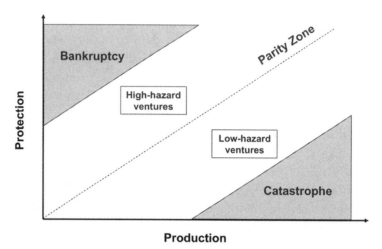

Production

Figure 1. Outline of the Relationship between Production and Protection[12]

Disasters occur when people in the organization inadvertently cross the boundaries after the executives and senior-level managers push too hard attempting to squeeze out the last few percentages of production. The top-level executives and senior-level managers focus their messages and attention on results and don't seem to attend to the quality of activities or processes. The emphasis from the top on results, with no attention devoted to the quality of operations, causes the middle-level managers, front-line supervisors, and operators to respond to the heavy influencing by taking shortcuts or accepting additional risk to achieve the results they perceive the executives are demanding.

One important characteristic of the relationship between production and protection can be seen in the notional chart. Notice the slope of the white parity zone. Even though the managers may never know precisely the current state of the organization relative to the boundary of catastrophe, one critical relationship can be known. Every time an executive or manager influences the personnel in the organization to increase production—without a simultaneous proportional improvement in safety—the organization likely will be driven closer to the boundary of catastrophe.

---

12   Reprinted by permission of the publishers; James Reason, "Figure 1-2 Outline of the relationship between production and protection," *Managing the Risks of Organizational Accidents* (Burlington, VT: Ashgate Publishing Company, 1997), 4.

To maintain the desired margin of safety, increases in production must always be accompanied by improvements in safety. The next time you hear an executive or a manager delivering a passionate speech about the need to improve results by increasing production, listen to the corresponding words regarding safety. See if you can infer how well the executive or manager understands the critical relationship between increasing production and improving safety. If he or she doesn't understand this critical interdependency and doesn't seem to push as passionately for improvements in safety, the organization may be headed for disaster.

Staying in the Zone

How do leaders ensure the organization does not breach the boundaries of catastrophe or bankruptcy, which may be impossible to define? The good news is leaders don't need to have explicit knowledge of the precise boundaries to avoid either disastrous result, indicated in Figure 1. Rather than pushing for increased production or results, which are outputs, leaders should encourage proper operating practices, which are inputs. Operating leaders use effective leadership skills to enable, motivate, and inspire their workforce to use the published policies and procedures, with the designed systems and processes, to perform high-quality operating practices. Over time, the high-quality practices create improved performance, which results in increased production.

I think the previous paragraph is worth summarizing and simplifying because this is one of the more important concepts for you and your organization as you conduct missions while controlling risk in your dangerous world.

What's the goal of your organization? The goal is to accomplish as much as possible using high-quality processes—with no accidents. *How do you do that?* You follow the standards and practices of operating excellence. *What are the standards?* They are the entire collection of written policies, rules, and procedures for operating. *What are the practices?* They are the established processes, techniques, methods, decisions you make, and actions you take to operate well. *Why does this work?* Here's the beauty and elegance of this concept. People with experience in operations developed the standards and practices using knowledge of previous accidents and financial problems in your industry. So the standards and practices inherently describe how to operate within the boundaries of catastrophe and bankruptcy. They also include the current constraints of operations, maintenance, engineering, regulations, the condition of the equipment, and readiness of the people. The standards and practices represent the collective wisdom of your organization on how best to accomplish operating tasks, activities, jobs, and roles, in service of the mission.

So what's the hard part? This may be the right question to address as you conduct missions and pursue objectives. Conforming to all the standards and practices of operating excellence is extremely challenging for everyone in your organization, from hired to retired. When you're new and inexperienced, there are too many standards and practices to hold in your head as you learn to operate. By the time you have sufficient experience to deeply understand the standards and practices, you're at the end of a long, productive, rewarding career filled with learning. Some of you in the middle think you can do better without following the standards and practices that have been written in blood. I don't need to explain how that will end. And finally, for those of you who are trying to do a good job, dedicating yourselves to the mission, and contributing to a noble cause—you are the heart and soul of your organization. Continue to learn, share your knowledge, develop skills, and display the right attitude by following and promoting the standards and practices that define high-quality processes and operating excellence to take your company higher.

The Characteristics of Operating Leaders in High-hazard Occupations

Good executives emphasize working in better ways to accomplish goals, contribute to society, and increase the long-term value of the organization. In hazardous occupations, the emphasis to work in better ways is vitally important in preventing a decremental drift toward the next accident. The success of the organization depends on the ability of its executives to use the proper leadership skills to inspire managers and operators to work together in a high-quality and safe way as they confront hazards.

Leaders in all businesses can learn from observing the behaviors and skills of those who lead people in dangerous environments where the penalty of failure can be death. In his insightful book, *In Extremis Leadership: Leading As If Your Life Depended On It,* Thomas Kolditz describes some characteristics displayed by leaders in hazardous occupations.[13]

In my experience, three characteristics stand above all others. Successful operating leaders consistently demonstrate *commitment, caring,* and *competence.*

Commitment

Operating leaders who are successful in dangerous occupations are deeply committed to pursuing a noble cause. These leaders inspire similar commitment in their teams to accomplish high goals in service of the mission. This personal commitment to the mission gives operating leaders and teams the courage to face hazards. The leaders and their teams work together pursuing a greater glory that transcends the desires or fears of any individuals on the teams. Dangerous environments are inherently

---

13   Thomas A. Kolditz, *In Extremis Leadership: Leading As If Your Life Depended On It* (San Francisco: John Wiley & Sons, Inc., 2007).

motivating, as leaders and teams must achieve operating excellence to stay alive and accomplish missions.

For anyone engaged in a hazardous endeavor, the mission is much more important than the money. Astronauts who fly in space do not receive extra salary or financial incentives. We received no hazardous duty pay. The mission commanders received no command-at-sea pay, even though we were commanding a crew in a government vessel deployed on the (really) high seas.

After each mission in space, I did receive some extra money, though. Because I was a military officer on official travel orders ("Itinerary: Houston, TX, to Kennedy Space Center, FL, to Earth orbit, to KSC, FL, and return to Houston, TX, with deviations authorized as mission requires"), I was given a small increase in pay, called per diem, which is normally provided to offset expenses. On mission STS-102, because the government provided transportation, meals, and lodging for our thirteen-day deployment to Earth orbit, and one week in crew quarters prior to the launch, the compensation amounted to $2.00 per day. Mileage expenses were added at the rate of 34.5 cents per mile for the twelve miles I had to drive in my personal car to the airport before departure, plus twelve more miles for the return trip because my wife had to drive our car home.

The total amount I received after my fifth mission in space was $46.28. With this extra money, I bought Happy Meals for my family after safely returning. The mission was never about money.

The deep personal commitment to a mission drives successful operating leaders and their teams to pursue operating excellence, which improves performance. Better performance allows the team to achieve more missions with higher quality results longer into the future.

Caring

In addition to their commitment to the mission, leaders who operate successfully in hazardous occupations care deeply for their people who are contributing to the mission. The leader can't accomplish much without a team, but the leader's care for the people originates from more than the need to come together to accomplish any particular goal. The best leaders receive personal satisfaction from helping members of their team actualize their greatest potential as contributing members.

As with the commitment to a mission, the personal care for the people gives the operating leader and teams the courage to face hazardous situations. The care for others is a special human trait. People are driven to face danger and accomplish incredible feats, while overcoming tremendous challenges, in an effort to protect teammates from harm and accomplish missions for the greater good of humanity.

Competence

Members of teams are willing to follow leaders, even into hazardous environments, only if the leaders are highly competent. Professional competence is one of the foundational skills required to build trust. When a leader possesses superior knowledge and elite skills, the team will trust their leader to help them survive, overcome obstacles, and accomplish the mission.

Deep competence comes from the ability and desire to learn. In a dangerous business with volatile hazards, building competence requires learning quickly, or the organization and its people will not survive. Leaders who operate successfully among hazards always tune their senses to the external environment. Operating leaders are unconcerned with self, their situation, or how the hazards may affect them physically.

The survival requirements of identifying hazards, learning quickly, and taking corrective actions to control the hazards drive the leaders to search for vulnerabilities continually. As soon as a hazard is identified, the best leaders create immediate learning across to other teams in the organization so corrective actions can be put in place to save additional lives. Leaders and teams operating in hazardous occupations learn quickly—or they die quickly.

Preventing Accidents

My experience in dangerous operations leads me to believe all accidents are preventable. In almost forty years, after seeing or learning about hundreds of accidents, I have never seen an accident that could not have been prevented. I will grant a few would have been extremely difficult to prevent had I been the victim. These are the accidents that have really intrigued me. These are the accidents that motivated me to develop and codify the *Techniques for Operating Excellence* I present in chapters 4 and 5.

Every potential accident gives signals before it becomes an accident. From an understanding of the conditions in place before previous accidents, the organization can learn to identify those signals and change the conditions to prevent potential accidents in the future.

Managers in the organization turn hindsight around and develop foresight. Through investigations, the organization can learn why decisions and actions made sense to people at the time before the accident. The organization can understand past and current vulnerabilities and search for future vulnerabilities based on anticipating the changing conditions, influences, and hazards.

With a deep understanding of the sociotechnical system and how leaders are influenced and exert influence, the leaders in all levels of the organization can provide appropriate guidance to operating personnel to enhance profitability while preventing decremental drift toward an accident.

To achieve company goals, top-level executives and senior-level managers need to emphasize long-term profitability. But they must allow some conservative slack in operations and not push for ultimate optimization and efficiency, which would unintentionally drive the organization closer to catastrophe. Middle-level and front-line leaders need to emphasize operating excellence through the proper inputs and processes, using conservative decision-making to increase long-term profitability. Taken together, there should be complementary guidance from all levels of the organization to make effective decisions and improve the quality of operations in the present to increase profitability in the future.

To enhance the viability of the organization in the long term, the top leaders need to create leaders for the future. The organization should help leaders at all levels develop and improve their decision-making and leadership skills through teaching, simulation, mentoring, and providing experiences that are progressively more challenging.

Conclusions—Organizational Management of Risk

In this first chapter, I have written about organizations and their challenges in managing risk. Organizations manage risk well most of the time. Yet they continue to be blindsided by accidents that were not only preventable but, to complete the frustration, had been experienced before. The managers were not bad people, and no organization intended to act unsafely. In their zeal to increase results, the top-level executives and middle-level managers pushed too hard to squeeze out the final one or two percent of increased performance, and they inadvertently crossed the line to disaster.

To survive in today's competitive environments, corporations must improve results and increase profitability to enhance their ability to contribute to society into the future. At the same time, organizations must prevent accidents to survive. Many managers seem to believe these requirements compete. Safety requires investments in time and money, which decrease the bottom line of profits in the short term.

Results are the most important objective for an organization. But managers can't get results by pushing for results, certainly not in the long term. After a quick burst of improvement in production attained with a push for results, accidents will follow, which will destroy any short-term improvement and kill people, if not the whole organization.

The solution to the seemingly paradoxical requirements of investing money to prevent accidents while generating more profits is not easy for a manager whose boss is demanding only profits. Executives and managers don't seem to realize the only variable the workers can change is the quality of their activities.

Rather than pushing directly for improvements in the bottom line of profits and results, the executives and managers should encourage the front-line leaders and workers to *improve the quality of their activities*. This makes sense, even with a cursory understanding of how the workforce conducts operations. Workers don't create results, individually or even collectively in the short term. Workers only conduct activities. Results are the summation of collective activities integrated over time.

Improving the quality of activities in operations simultaneously prevents accidents and achieves the executives' desired goal of increased profits. This is not a coincidence. Improvements in quality yield improvements in safety and profitability because workers are operating in better ways. With a deep understanding of how this principle works on the front lines of hazardous operations, executives and managers will achieve the results they want and enhance the viability of their corporation in the long term.

How should managers encourage operators to improve the quality of their activities? Almost always, managers focus on the *technical side* of the sociotechnical system that is the organization. Working on this technical systems side seems easier for rational managers who learned to be successful in Western business cultures. But managers who are more skilled on the *social side* are much more effective in influencing the workers to improve the quality of activities. Though harder to master, this social side can be powerful. Remember, systems don't launch people into space. People use systems to launch people into space.

Here is another way to appreciate the value of the two sides in the sociotechnical system. NASA needed managers who were good at the systems side to send astronauts to the moon. But we needed leaders who were skilled on the people side to work together and bring the Apollo 13 astronauts home after they suffered an explosion on their way to the moon. How to emphasize the people side is the focus of the following chapters in this book.

Before we leave the systems side, here are a few concluding thoughts on how organizations manage risk, usually effectively, but occasionally with disastrous results. History indicates preventing every potential accident by managing risk through organizational processes alone is nearly impossible. Preventing accidents requires exerting control over risk. To reduce risk effectively and completely, two kinds of control must be exerted: *organizational control* and *personal control.*

Organizational control of risk can be thought of as managers and workers collectively using systems to reduce risk to the organization as a whole. Hazards sensed and identified by individuals or groups are brought centrally to a manager or decision-making body for determination of the proper corrective actions that will reduce risk and improve the performance in the collective organization. Most organizations devote the bulk of their attention to this kind of control. Organizational control can be effective but is often insufficient in a dangerous business without front-line team leaders and operators excelling at personal control.

Personal control of risk is exerted by individual leaders and workers who are personally making decisions and taking actions to prevent accidents and improve performance based on what they sense in the moment. Personal control, when it supplements organizational control, has the power to eliminate all accidents, improve performance, and take the company to new heights. Organizations tend to devote insufficient effort to teaching, training, encouraging, and incentivizing people to develop and hone the necessary skills to exert personal control. Personal control of risk will be the focus of the remainder of this book.

In summary, both organizational control and personal control are required. Organizational control methods are systematic directives developed and implemented by the leaders in the organization to help the workforce conduct their jobs effectively in a safe and productive manner. These methods are usually included in a *Safety Management System* in the organization. A simple list of the organizational control methods includes, but is not limited to, the following items:

Organizational Management of Risk

- **Safety Management System** (rules, policies, and procedures for the entire organization)
- **Local Safety Management System** and **Organizational Practices** (for different local areas or entities in the organization)
- **Managing Risk** (Process Safety)
  1. Identify the Risks (and Hazards)
  2. Assess the Risks
  3. Respond to the Risks (Implement Controls)
  4. Monitor and Review the Risks (and Controls)
- **Process Safety Methods**
  o Control of Work (Permitting Process, Job Safety Analyses, etc.)
  o Hazard Identification and Evaluation
  o Risk Assessment
- **Hierarchy of Controls** (Hazard and Error Defenses)
  A. System Design (Hard Defenses)
     1. Design Out (some organizations use Elimination or Redesign)
     2. Engineer In (some organizations use Substitution or Redesign)
     3. Guard Against (some organizations use Isolation)
     4. Constraints and Affordances (Human Factors design characteristics)

    B. Administrative Rules and Procedures (Soft Defenses)
        5. Warnings
        6. Training
        7. Personal Protective Equipment

Personal control methods include the techniques used by individuals to keep themselves, teammates, and equipment safe during work. These techniques are presented in the fourth and fifth chapters of this book. When supplementing the effectiveness of the organizational control methods, these personal control methods can help people reduce injuries, eliminate fatalities, conduct work effectively, and achieve more missions with higher quality. An operator can use these powerful control methods to improve safety. Using personal control methods requires the development and refinement of several skills. I will now turn to the personal side of controlling risk.

Chapter 2

CONTROLLING RISK
An Operator's Perspective

In the first chapter, I presented some thoughts on how managers in organizations manage risk, or at least attempt to manage risk, and how they influence the workers to improve performance and increase productivity. For the most part, managers do a credible job of reducing risk and providing good processes for operators to follow. But to stay alive in dangerous environments, individual operators cannot and should not rely solely on the collective organizational processes, even in organizations with the noblest intentions.

The front-line operators must face hazards and make life-and-death decisions in dynamic and complex operational situations. They use specific organizational processes and procedures that may or may not be applicable as soon as they are published. The operating environment is always changing and the operating systems are often deteriorating. Each operator represents the last line of defense in the organization intended to contain the danger and prevent the tragedies of death and destruction.

For thirty-five years, I have had the privilege of working as an operator and leading other operators in pursuit of important missions in hazardous environments.

27

This has been my passion. It is not the risk I enjoy, and I don't like the feeling of adrenaline, resulting from perceived danger. Rather, what drives me is my desire to accomplish noble missions while helping my team coolly and mindfully face the hazards and control the risk with operating precision.

Acquiring the skills to be successful in dangerous environments is a continuous journey. I have benefited from many other experienced operators and inspirational leaders. In appreciation, I hope to share what I have learned to help others save lives and perform valuable missions under hazardous conditions.

The Differences Between Managing Risk and Controlling Risk

To predict and prevent the next accident and improve performance and productivity in hazardous environments, organizations must master the processes of managing risk and controlling risk. Both are important. Both are necessary to ensure long-term health and viability of the organization, allowing it to contribute valuable goods and services to society into the future.

Through my experiences, I have learned to appreciate at least three important distinctions between managing risk and controlling risk. I will summarize the differences briefly and will add more details later.

First and most importantly, managers view risk as a notional calculation of unwanted outcomes. To the operators, risk is not a notional concept, but an actual condition of real hazards that must be controlled to stay alive. I don't mean to infer a judgmental distinction between the two viewpoints of notional and actual risk. Both perspectives are valuable. When managers conceptualize the various organizational risks from their offices, they can calculate and quantify the risks in a relative way and can make decisions to reduce the risks to the benefit of the collective organization and the individual operators in the field. When operators see real hazards, they take real actions to eliminate or reduce exposure to those hazards to protect themselves and the organization.

The options available for reducing risk are the second major difference. Managers can make changes to the system and the equipment to reduce risk to an acceptable level. Operators can't. Operators must play the hand they are dealt. They must use the system and equipment as designed and delivered. This difference between managers and operators is parallel to that of test pilots and operational pilots in the fleet. As a US Navy test pilot, I was responsible for identifying deficiencies in the aircraft or systems and making recommendations for corrective actions to eliminate or mitigate the deficiencies. As an operational pilot deployed at sea, I had to accept the design deficiencies in the airplane, no matter how bad they were. Of course I followed the mandated procedures intended to keep me safe, but sometimes the procedures weren't good enough. If I wanted to stay alive—and accomplish the mission—I had to deeply understand the design deficiencies, my personal capabilities, and how to use the system in the best way possible, while still following the prescribed rules and procedures that might not have been optimum.

These procedures lead me to the third important difference between managing risk and controlling risk: the actions managers specify the operators *must* take to manage risk, and the discretionary actions the operators *should* take to control risk. A common product of the risk management process is the issuance of rules-based procedures. When the managers evaluate the level of risk in the organization is too high, they make rational technical decisions to reduce that risk to an acceptable level or lower. After the system design has been finalized, the next mechanism the managers use to limit the risk is mandating the use of rules-based procedures.

To operators on the front lines of danger, these prescribed procedures are necessary but insufficient. When the operators follow the procedures, the risk is kept at or below the organizationally defined acceptable level. But remember, the operator is trying to survive and be successful in a dangerous world that is not so logical and defined. The hazards are always present, even when the manager has calculated the risk is acceptable. The systems are always degrading, even after the rules-based procedures have been prescribed. So the operators need something more to control risk, some discretionary techniques to prevent *all* accidents. The answer is principles-based techniques, and I'll get to that soon.

Here are a few more details about the three differences between managing risk and controlling risk, which are, (1) the notional and actual concepts of risk, (2) the options available to reduce risk, and (3) the prescribed procedures for managing risk supplemented with discretionary techniques for controlling risk.

Risk management is the organizational process managers use to reduce the collective risks in an organization to acceptable levels or below. From the perspective of the managers, risk is notionally calculated as the mathematical product of two parameters: the predicted likelihood of a hazard manifesting as an event with bad consequences times the severity of the consequences if that event happens.

$$Risk = Likelihood \times Consequence$$

This concept of risk, which is common in organizations, guides managers in making decisions appropriate for managing the collective risk in the organization. Using the formula for risk shown above, managers can reduce the calculated risk by taking actions that reduce either the likelihood or severity of consequences, or both.

Managers use two mechanisms to reduce the likelihood or the consequence. They either change the equipment or systems, or change the ways the equipment or systems are being used. Managers can add, modify, or delete hardware or software.

They can mandate changes in the ways the operators use the systems. They can put physical or procedural barriers in place to prevent operators from using the existing equipment and systems in certain ways. To document and promulgate any of these changes, managers issue updated standards (specifying the design of the operating systems) and procedures (specifying how the operators use the systems).

Operators think about risk differently than managers. At the front line of operations, individual operators face the danger every day, moment to moment. Operators view the risk as simply the hazard, or the danger, and the hazard is an actual condition, not a notional calculation. The hazards, all of them, must be dealt with successfully, in the present. The operator does not have the option to redesign or change the hardware or software. If the operator is to be successful and have the opportunity to conduct operations in the future by staying alive today, each task must be completed safely and successfully using the operating system the organization provides.

The likelihood of a hazard becoming a bad event, as evaluated by a manager, is somewhat irrelevant to an operator, who cares only that the bad event doesn't happen.

The operator's view of the magnitude of a consequence is quite different also. For example, managers might evaluate the severity of a potential consequence as two major injuries throughout an organization with 5,000 people. This value is not important to an individual operator who is simply trying not to be injured during hazardous operations on any particular day.

So, managing risk in organizations is different than controlling risk at the operator level. Both processes are necessary and valuable. Managers can take actions to reduce the risk to an acceptable level. Managers can change the equipment or the system if the risk is too high. The operators don't have that luxury. An operator must use the equipment and the system as they exist on the day of operations, once the risk is deemed acceptable by the managers. To control the risk and prevent injuries and catastrophes an operator's only available option is to use the equipment and the systems wisely. Sharing and explaining how operators should do this is the primary theme in this book. For now, I will continue to describe the operator's perspective of controlling risk.

Accident Prediction

Accidents give signals before they become accidents. Some signals are obvious and demand attention. If any observers are attuned to these signals, a potential accident can be averted. Other signals are subtle and ephemeral. Only the skilled—or lucky— observers can identify the signals and allow operators to take appropriate actions in time to avoid the accident. But, one sad conclusion can be drawn regarding every accident in the history of the world that has resulted in minor injuries and damage, or major tragedies and disasters. Before every accident, either the signals were not identified, or effective actions were not taken to prevent the accident. In this sense, the impending accident was unknown to the victim.

A few years ago, I had a conversation with managers in a large company. I was discussing the need to identify the signals of impending accidents. Eventually, a manager asked me the critical question that had concerned them in their hazardous endeavors: "How do we uncover the unknown if we don't know where to look or what to look for?" I told her this is the most important question for leaders to address if they want to be successful in high-risk operations. Developing ways to uncover unknowns is what has kept me alive during four decades in dangerous operations. In developing these ways, I learned to supplement rules with principles.

Rules-based Procedures versus Principles-based Techniques

With the knowledge of what the signals looked like before previous accidents, managers can promulgate new rules and procedures to help operators prevent future accidents. Companies usually default to this simple type of preventative safety by specifying *rules-based procedures* and mandating their use.

But, rules and procedures are often closed and nonadaptive. They apply to specific operations or conditions. What if the conditions change? What if the processes are different for a new operation on a new day? Operators and managers won't always know what new signals the next accident is starting to send.

Unknown future accidents are best identified and prevented using discretionary techniques for operating excellence based on comprehensive principles of operations. *Principles-based techniques* apply to many more situations, including ones that may have unique conditions for which no rule has been specified.

Over the years, I have observed that teams and companies don't usually teach these valuable open and adaptive techniques based on principles. The absence of this kind of teaching and operating in organizations likely comes from one or more of four situations:

- Managers may not understand the power of open and adaptive principles-based techniques to prevent accidents and improve performance;
- Managers may not believe their workers can or should be trusted to operate reliably with *discretionary* open and adaptive principles-based techniques;
- Managers may believe principles-based techniques are too difficult to develop, teach, understand, and use;
- Managers may believe better performance will be achieved simply by mandating compliance with rigorous, closed, and nonadaptive rules-based procedures.

Managers who don't understand the power of using principles-based techniques to prevent accidents likely haven't had relevant operating experiences in high-hazard situations. Overcoming danger and avoiding death and destruction are personal experiences that would help individual managers appreciate the benefits of the discretionary techniques for operating excellence based on comprehensive principles of operations. The best operators use principles-based techniques not only to prevent accidents but also to improve performance and increase long-term value in the organization.

Example: Cell Phones and Driving
A quick example may illustrate the relative merits of rules-based procedures and principles-based techniques. Consider the cell phone and driving. What's the rule? Many companies prohibit the use of cell phones while driving on company time. Drivers who follow rules-based procedures will likely prevent many motor vehicle accidents.

But if you want to prevent *all* accidents, what's a wider principle you might follow? Don't be distracted. Written positively, maintain awareness of your driving situation. Principles-based techniques for driving defensively might include recognizing and avoiding *all* potential distractions, observing all hazards, anticipating the actions of other drivers, and predicting and avoiding dangerous situations. Drivers who follow principles-based techniques have a higher likelihood of predicting and preventing all accidents while driving.

Identifying accidents coming at operators from the *unknown* requires developing and mastering *principles-based techniques*. Principles, rather than rules, are open, adaptive, and comprehensive. The principles-based *Techniques for Operating Excellence* (illustrated in chapters 4 and 5) are the result of what I learned, developed, and used in Flight Crew Operations at NASA to help our operational teams become more successful. My appreciation of the power of these techniques comes from a lifetime of thinking about operations, facing hazards, and controlling risk in dangerous environments, beginning at an early age.

My Early Interest in Flying
I don't remember much before I was ten years old. Only a handful of situations registered in my memory. I must have been merely acting and reacting as I passed through my environments in day-to-day life. I imagine this is what the world must seem to animals, with no connection from the past to predictions of the future

and, therefore, no understanding of the relationships between risk and long-term consequences.

Sometime when I was ten, a main circuit breaker engaged in my mind, without a specific causal event, and I began to remember—not only isolated incidents but also entire days, including mundane events. At the time, my brain seemed able to sense, assess, interpret, and record everything around me. The world suddenly became an interesting place, and it seemed to hold infinite possibilities. I had a great family, and I began to notice the positive effect they had on me. In my sphere as an introvert, I used my mental bandwidth to capture as much input as possible; I had little interest in broadcasting.

Above all this processing of newfound information, I was quite interested in *how* my mind was thinking. In this meta-level processing, I noticed a specific and certain destiny about my future. This realization wasn't overwhelming or dominating; it was just there. The belief of what I would do later in life came from a high confidence in my ability to enter dangerous situations and control both my internal thought processes and external control systems.

I wasn't as physically skilled as my older brother, who was the best athlete I ever saw in a wide field of neighborhood athletes. But nonetheless, I knew what I would be good at, and I had a great desire to control vehicles, especially ones that could overcome gravity. Just as I felt my brother would become a professional athlete (he later became a Professional Golf Association card-carrying pro), I knew my future would involve my ability to control anything in dangerous operations.

So in 1962, my ten-year-old, naïve mind held no doubt I would become an astronaut. Later, as I grew older, I realized my chances were infinitesimal, but I never dwelled on the realization I might not have the chance to be an astronaut. I was content in my belief I was born to fly, and I was put on this planet to leave it (and come back). If offered the privilege, I would work hard to master the art and science of controlling space vehicles.

Personal Risk Attitude

Even at a young age, I began to think about danger because I knew it was a necessary part of my desire to fly. My initial thoughts about risk were essentially the same ones I have today: I don't like danger, and I do what I can to avoid it. When I choose to enter a dangerous environment because I desire to achieve an objective, I attempt to minimize the operating risk by exerting control over the hazards. The greater the risk, the greater the control.

I appreciate why engineers, economists, and executives evaluate risk as the product of likelihood and consequence. This numerical calculation gives the managers a rational method to compare relative risks and make informed decisions as they prioritize which risks to work on and how much money and time to spend on the risk-reduction efforts. The calculation can make the process of risk management more consistent across the organization even though disparate risks in distributed systems are calculated by different managers vying for funding from a centralized bank in the wider organization.

But there is a downside to using this rational, rigorous, technical method for calculating predicted risk in an organization: it's not always accurate. Humans don't have a good feel for the uncertainties in complex sociotechnical systems when they try to predict probabilities.

Small uncertainties have large effects on probabilities calculated near one or zero, representing near certainty or near impossibility. Unfortunately, many people think they do understand high and low probabilities, which gives managers misplaced confidence in their ability to make good decisions regarding risk.

NASA engineers and managers could calculate risk to seven decimal places, but the inputs to their calculations had high uncertainty and were sometimes based on gut feel and good ol' Kentucky windage. Garbage in, garbage out. A kinder way to describe the risk assessment methods might be to say the managers didn't always understand the uncertainties in some of their assumptions.

When I joined the space program before the *Challenger* accident, the predicted probability of loss of vehicle after making some improvements to the engines was evaluated to be about 1 in 10,000. This was far from the result observed from flight history at the end of the program, which was 2 losses in 135 flights.

Evaluating personal risk using a rational calculation of likelihood times consequence, as an organization does, is not useful to me.

First, consider the factor of likelihood. The predicted frequency or probability calculated by the organization of sustaining a bad outcome matters not to me as an individual operator. If I fail to control the hazard and I suffer a broken leg, the predicted probability of that same event happening over a statistically significant number of operations is irrelevant. My leg has broken with certainty.

Next, consider the calculation of the severity of the consequence. In a dangerous example, the managers might calculate a certain hazard could result in "one fatality" over ten years of operations, and another hazard could result in a more severe consequence to the organization of "three to six fatalities" over the same ten years. An evaluation of the relative severity of consequence of the two hazards is meaningless to me if I am one of the three to six who is killed.

So how do I assess risk and decide whether to confront the danger I fear?

Rather than using rational, rigorous, technical calculations, I use experience, judgment, and human values in my decision-making algorithm. Before I decide to attempt any activity in a hazardous environment, I consider three interrelated aspects of the situation related to risk:

1. The worst *consequence* if I fail to control the hazards
2. The *criticality*, or importance, of the mission my activity is supporting
3. My ability to *control* the hazards

Consequence, criticality, and control are the factors that matter to me when I decide to accept the risk.

Consequence

I start by assessing the possible consequences. Of these, I always assume the worst consequence will happen if I don't execute the operation with exceptional control. I don't consider the predicted probabilities of the consequences, especially if unlikely, because too many factors in the real world can invalidate the predictions for dangerous operations. The price of entry into the hazardous environment is my willingness to withstand the worst consequence to the organization and me. If I can't pay that price, because the worst consequence is too severe, I won't attempt the activity.

Criticality

Before I walk away from an operation with a potentially severe consequence, though, I consider the importance of the mission. The greater the importance of the mission, the higher my tolerance becomes for accepting a severe-level worst consequence. When the mission is critical, my increased willingness to accept a more severe worst consequence, and thus face greater risk, is always supported by a greater effort to perceive and control the hazards. I can't allow failure in a critical mission.

Control

Finally, I consider the third aspect of the risk situation. Even in highly dangerous missions, the best way to reduce risk is to control the hazards. If I can control the hazards down to the level of consequence I'm willing to accept, I attempt the mission. If I can't, I don't.

Once I decide to accept the risk and I begin to prosecute the mission, these three considerations collapse to one. From then on, it's all about control with operating excellence. As an operator, I must control the hazards. The more important the mission, the greater the control I exert, for one reason: I don't want to fail.

When I decided to accept the mission assignment, I was willing—and expected—to suffer the worst consequence if I didn't control the hazards. But once I engage in the operation, I control the hazards as best I can to prevent that consequence from occurring. Worse than mission failure, I may not be asked to attempt more missions if I don't control the hazards. I certainly won't be asked if I'm dead.

Even in a less dangerous mission, when the likelihood of a bad outcome is remote, my desire to control the hazards is always present when conducting an important mission. Probabilities don't matter. Control creates success. No matter how high the danger or how probable the worst consequence, I can always reduce the risk if I control the hazards.

If my method for assessing personal risk seems too involved, here are some quick examples to illustrate how I make decisions about risk using the concepts of consequence, criticality, and control.

Example: Gambling for Money

For me, gambling is not a critical mission. On the rare occasions I have gambled, I always expected to experience the worst consequence. I only risked the amount of money I was willing to lose, and I expected to lose it all. When gambling, the only reliable control method available is quitting. In all four experiences on gambling trips, I have left town on the plus side of the ledger, with net winnings of $55, $1.35, and $0.05 in Las Vegas and £20 in London. (When I was twelve years old, my parents let me play Vegas nickel-slots; in 1964 the hotels had little regard for age, and I stopped as soon as I was five cents up while learning how to control risk.) Risk is about control, and when so little is available, control must be executed with dispassionate precision. In the case of gambling—quit while ahead.

Example: Riding a Motorcycle

Because I love controlling systems and equipment, I wanted to experience controlling a motorcycle. One time. My mission was to have the experience. The worst consequence was dying, but only if I allowed my mind to be distracted and didn't execute the proper control. Death is a severe consequence. To reduce that consequence to an acceptable level and stay alive, I would need to exert flawless control with the proper awareness of my surroundings. If I didn't, I expected to die. So I accepted my personal mission and decided to control a motorcycle with operating precision.

On a morning with good weather, in two trips covering twenty miles, I exerted the precise control required to prevent death, or even eating a bug. After I was successful, my operation to experience controlling a motorcycle was complete. I haven't been on one since, because the low criticality for me of riding a motorcycle no longer increases my tolerance for accepting the worst consequence, and it is too much effort to exert the control required to stay alive.

Example: Sledding on Snow

Here is an example with a less severe potential consequence that was nevertheless unacceptable. In the winter of 1980, I had an opportunity to go sledding on a ski hill with friends. As I assessed the situation, the worst physical consequence could have been a broken arm or leg. At that time in my career, I could not tolerate this outcome, and I couldn't exert sufficient control over the uncertain environmental hazards to prevent this outcome. In the face of great peer pressure, I refused to go down the hill.

The problem wasn't fear or potential pain or my ability to control a sled. The issue was that I had recently been accepted to the US Naval Test Pilot School, and I refused to risk a possible delay in my class start date. That was the more relevant possible outcome. Simple decision. Expect the worst consequence (unacceptable delay in start date) and evaluate the criticality of the sledding mission (extremely low). No need to go any further and assess my ability to control (too many environmental uncertainties). Mission denied.

Example: Travelling on a Commercial Airliner

If you are wondering how I feel about riding on a commercial airliner, here are the two relevant parts of my limited assessment. My fellow passengers and I have little control over the outcome of a flight. What control I have, I am prepared to execute flawlessly every time I board an airplane. For example, I count the number of rows between my seat and the emergency exits, in both directions, so I can find either exit with no visibility. Keeping my eyes open in a smoke-filled cabin will be difficult. To enhance my preparation, I attended a valuable training course United Airlines offered and experienced egressing a dark, smoky cabin with almost no visibility. I'm as prepared as I can be.

But now consider this. Riding on a commercial airliner is not a dangerous operation—not even close—so I don't need to invoke my personal risk algorithm by evaluating consequence, criticality, and control. In this case, statistics are useful because they are based on historical data, not predictions. In any given year, my odds of boarding a US airliner that will have at least one fatality are less frequent than 1 time in 4,000,000 flights. Again, most people don't have a good feel for such large numbers in the denominator. To demonstrate the rarity of those odds, I must flip a coin twenty-two times in a row and have it come up tails—all twenty-two times! If a single head comes up in any of those flips, I might as well relax on the flight, as today is not the day to die.

But I still count the rows to the exit.

Aviation Operations in the US Navy

I love the feeling of control. I was drawn to the challenge of risk but only in the sense of controlling risk. I never wanted a "real" job. I only wanted to fly.

I joined the US Navy to master the challenge of landing on the pitching deck of an aircraft carrier, at night, in a thunderstorm. Though the danger might be high, I knew the resultant risk would be low if I could control several tons of high-rpm, screaming hardware, while also keeping my thoughts and emotions under control.

Though I was a little worried about having no previous flight experience when I entered flight training, I learned the Navy preferred prospective pilots with no experience. They said it was because we were impressionable and had no bad habits to break. As expected in the military, the instructors began with the basics and taught us the Navy way to fly. Everything was oriented toward landing on an aircraft carrier.

I still remember the powerful impression left on me in my first week of ground school training for jets in 1977. The instructor casually explained the insidious concept of vertigo, how to identify its symptoms, and how to combat the strong, yet erroneous, sensations that might lead a pilot to fly a shiny airplane into the ground. The key was to use the airplane's instruments and ignore the sensations from the neurovestibular system, which was designed through millions of years of evolution while shackled to the earth by gravity. The instructor explained it will be difficult to ignore all that history, but to stay alive you must trust your instruments.

As I sat listening, my first thought was, *Well, this is scary. I had better pay attention. Everything they are teaching me might save my life someday.* My second thought was, *What if the instruments are wrong at the same time I experience vertigo?* Even then, without realizing it, I began to formulate the concept of using *dissimilar verification,* one of the techniques for operating excellence we developed later as astronauts. One instrument might be wrong, but they won't all be wrong at the same time in the same way if they are independent. I must figure out how to cross-check the instruments and determine which ones are right. Not easy, but the payoff would be substantial.

Years later, when I experienced vertigo for the first time on a night with a weird cloud formation and no visible horizon, I thought it was intriguing. I immediately flashed back to the advice of the instructor and trusted (after verifying) my instruments. Vertigo was not so fun for a squadron mate who was trying to land on our aircraft carrier that he sensed was steaming through the ocean in a thirty-degree angle of roll. The landing wasn't his best, but with the proper training to ignore his faulty sensations, he survived. It is too bad JFK Jr., the only son of the namesake of our aircraft carrier, the USS *John F Kennedy*, didn't have the same naval aviation training to control risk before he failed to survive in the vertigo-inducing conditions near Martha's Vineyard in July 1999.

In my initial flight training, I began to develop another valuable attitude about risk. As good as I thought the military was in training pilots, I knew they would not be able to teach me everything I needed to stay alive. The organization and I were partners in my survival, but I held at least fifty-one percent of the corporate vote. I needed to train more than they were offering. When I heard I would receive only six sessions in the simulator before I was expected to pilot a real jet for the first time, I knew six would not be enough to satisfy me. So I became friendly with the night-shift janitor in the simulator building. For the next week, he unlocked the door at one o'clock in the morning, and I practiced surreptitiously in the simulator. By the time I flew my first hop in the jet, I had six times as many hours as my classmates. After that first flight, my instructor asked me to confirm I really had no previous flight time.

For the next two years, the Navy followed the same pattern for training. They didn't waste time. They told us once, let us practice once, then threw us into the breach, solo, with an external safety observer. At any sign of weakness, we had remedial training, but who wants that? So I gave myself all the remedial training I needed with extra study, extra simulators, and extra cockpit rehearsals in the hangar at night. All's fair in love and training for war.

A naval aviator's first arrested landing on a carrier at sea is performed solo. I thought the philosophy was interesting. With an operation so dangerous, why risk the additional life of an instructor pilot?

Pilots aren't the only naval personnel who are exposed to risk every day. Life aboard an aircraft carrier is inherently dangerous. In 1979, I saw one of the most inspirational events in my short twenty-six-year life. I was in the ship's tower observing flight-deck operations. Suddenly a fire broke out on the deck. I saw five enlisted sailors sprinting toward the fire in attack mode. In civilian life, we are trained to run away from fires. At sea, there is nowhere to run. They were fighting for our home.

Photo 1: Port side of aircraft carrier, USS *John F Kennedy*

On a flight deck, the hazards are always around but not always obvious. There are invisible spinning propellers, jet engine intakes with an appetite to swallow unobservant sailors, and imperceptible exhaust ready to blow anyone over the side, regardless of rank. At night, it's dark and even more dangerous. If I tripped and slipped overboard, I would become a small dinner in a large sea, not likely missed and never found. Photo 1 shows the port side of our aircraft carrier.

During flight operations one day, I had just completed my engine start and was waiting for taxi and launch sequencing. Right in front of my cockpit, I saw a sailor being blown like tumbleweed directly toward the fantail in the high winds boosted by jet exhaust. With no fence or barrier to save him, he was desperately trying to gain purchase on the flat flight deck. I was powerless to help.

Another sailor calculated their mutual paths and executed a perfect intersecting trajectory to tackle the out-of-control sailor just before they both flew over the ramp to Davy Jones' Locker. The second sailor had a bigger wingspan and was able to spread out on top of the first and save both lives. I nominated the second sailor for a medal. He was awarded none, as life-saving behaviors are routine aboard an aircraft carrier. I don't expect the second sailor cared about the medal. He saw a crewmember in jeopardy and reacted to save his life.

Operating in Hazardous Environments

Dealing with risk is a dangerous endeavor. Successfully controlling the danger requires three elements: knowledge, skill, and the proper attitude about risk. Knowledge is the easiest of the three to specify, teach, acquire, and test. Skill in controlling risk is more complicated to describe, develop, and maintain. Skill can be improved through practice in real environments or simulators. More experienced operators and instructors can assess the safety skills of another operator by observing a demonstration of risk control in a dangerous environment, real or simulated.

Risk attitude is far more important than knowledge and skill. Unfortunately, attitude is also the most difficult to teach and to evaluate. Most operating organizations teach knowledge to their operators and train them to develop skills. Few organizations seem to devote direct effort to helping personnel develop the proper mental attitudes to control risk. In chapters 4 and 5, I offer powerful *Techniques for Operating Excellence*, which can be described, taught, learned, observed, and evaluated. These techniques are essentially mental attitudes that will help operators control risk and improve performance.

The best operators with extensive experience in dangerous situations have the best attitude about risk. These operators are not concerned with personal consequences or distinction, nor are they worried about internal emotions or stressors. The senses of these elite operators are directed exclusively toward external hazards in the operating environment. When they detect any hazard, no matter how small, they immediately analyze the threat, develop and assess a potential control measure, and immediately take corrective action to eliminate or minimize exposure to the hazard and control the risk. Almost as quickly, these operators share the valuable learning with their teammates and other teams that may face similar hazards.

My attitude about risk was shaped primarily by the experiences I had as I passed four significant milestones in my career. These were not typical milestones you might expect, such as the tragedy of a colleague's death or the elation of selection into the Astronaut Office. Rather, these were events that seemed inconsequential one minute and became life-altering the next. An observer, untrained in identifying the subtleties of mental attitude, might have seen no immediate differences in my demeanor or my behaviors. The profound change occurred between my ears. Each event caused me to modify and improve my risk attitude, which has continued to shape my way of operating in dangerous environments, even to this day.

The first of the four milestone events occurred in the Mediterranean Sea, while I was flying with Air Wing One off the USS *John F Kennedy*, in 1979. Photo 2 is a picture I took of my two wingmen while flying at low altitude on such a flight. The next three events occurred between 1984 and 1990, when I was flying with NASA. I will describe each of these four instances and how they crystalized and strengthened my attitude about risk.

Photo 2: A-7 Corsair, just off the western coast of Sardinia in 1979

Night landings at sea were interesting. When someone asks a naval aviator how many night traps he or she has made, a typical answer might be, "125—which one do you want to hear about?" That's because they are all different. And each one is seared into our memories with a psychological branding iron.

I was young and cocky, and I enjoyed every landing. Until one particular night. Early in my career, as is the situation with so many young men in dangerous professions, my confidence had begun to outrun my capabilities. On this one night, I would suddenly realize I was not as good a pilot as I thought I was.

Stories of flying around the ship often start with, "It was a dark and stormy night." Figure 2 is an illustration of what a typical approach might look like on a short final to a ship at sea, with no "Commander's moon" to illuminate the surroundings. Only the fearless (through inexperience) junior officers love this kind of environment.

Figure 2: Illustration of what a typical night landing might look like on a ship at sea

I was flying at 15,000 feet in a holding pattern, waiting for my push time to commence the approach with the proper one-minute sequencing, which was desired between aircraft in the landing zone on the ship. I could hear the other pilots ahead of me, their voices infused with increasing anxiety on the radio, as more and more of them "boltered," after missing all the wires. This indicated the ship was moving up and down in the sea swells a little more than usual, and the pilots were overflying all four arresting cables on the flight deck. Exactly the kind of challenge I liked. I was in my element, born to fly. As I commenced my approach, I was thinking about how cool it was, to be a naval aviator, doing what I do best. My brother had an average job back in New York. I had the best job on the planet.

Confidence is good. Overconfidence is dangerous. When I should have been concentrating on the task, I was thinking about the wrong things, and it almost killed me. I realized this too late, at about the time I crossed the ramp to a hard landing.

Fortunately, the damage was not to the hardware but only to my psyche, which was fairly well destroyed. As I taxied clear of the landing zone, I couldn't stop shaking. After shutting down in the darkest parking position on the flight deck, I didn't unstrap for about ten minutes. I sat in my ejection seat, frozen in the blackness, too scared to move. In those moments, I vowed to never again let my concentration waver during operations. If I wanted to survive, I could never allow extraneous thoughts to enter my mind and decrease my ability to sense and react, immediately and effectively, to the hazards always trying to kill me.

This was a lesson that would benefit me for the rest of my life. Not all hazards are in the external environment. *Some hazards are within.* In a dangerous situation, I became distracted and lost focus. The only acceptable solution was to master the ability to concentrate and never allow distractions to diminish my operating performance. For the rest of my professional life—in the US Navy, at NASA, in the oil-and-gas industry, and even on stage while playing drums in a band—I have continued my journey of developing mental discipline, attempting to master the skill of concentrating on the operational task and staying in the moment. This multiyear effort resulted in developing one of the Techniques for Operating Excellence (presented in chapter 5), *Be Mindful During Operations*, which has helped me succeed on countless occasions.

Spaceflight Operations at NASA

The prevalent emotions I experienced as an astronaut for twenty years were professional fulfillment with sustained excitement over the privilege of doing the only thing I ever wanted to do since I was ten years old. On the opposite side of the emotional spectrum, three sobering instances of self-reflection took place when the reality of the job hit me. These were the next three of the four milestones that profoundly shaped my attitude about risk.

The second milestone occurred on a tour of the Kennedy Space Center during our initial astronaut candidate year. As we walked around the vehicle in the Orbiter Processing Facility, I had the opportunity to crawl into the aft engine compartment of the Space Shuttle. The sight stopped me cold.

Leading into the fuel manifold for the three main engines was the largest fuel line I had ever seen, at seventeen inches in diameter. In an airplane, many errors are recoverable. In a Space Shuttle, with massive fuel lines that feed high-pressure turbo pumps running at more than 30,000 rpm, a single mistake can cause cavitation of the pumps, resulting in overspeed and mechanical failure of the blades that will rip through structure and explode the back end off the vehicle. I forced a deep breath, and I knew I needed to study and learn more than what the organization was capable of giving me.

The third cause for self-reflection and change in risk attitude occurred when I was five months away from my first flight in space. I was beginning to wonder whether I had what it takes to fly a "jet rocket," as it is designated in our aviator's logbook. I had logged 345 arrested landings on aircraft carriers, but this was different. One small invisible mistake in the orbiter could lead to one large visible fireball on the launch pad. On this one night, I learned about handling the pressure of spaceflight operations in an isolated, darkened facility known as the Software Avionics Integration Laboratory, or SAIL for short. Our job in this laboratory was to verify the integrity of the software resident in the flight control computers for an upcoming Space Shuttle mission.

During this particular test, I was the sole crewmember in the cockpit of the simulator. My task was to perform four hours' worth of procedures culminating in the mission-critical deorbit burn to commence reentry into Earth's atmosphere at a precise second in the timeline. In the laboratory, twenty-five engineers at consoles recorded data from real flight software loaded in the simulated orbiter's computers.

Several times during the evening we experienced problems in the simulation that delayed our progress. The emotional tension among the engineers was high as the test dragged on past midnight. In the cockpit, as the clock counted down to the scheduled time of ignition, I was getting further and further behind in completing the operational tasks normally accomplished by four crewmembers on orbit.

With two minutes to go, I realized the engineers would be furious if I failed to complete the burn successfully and we had to repeat the whole process the next night. I worked as fast as I could to complete the procedures as written with thirty seconds to spare. Fourteen years of flight experience told me that mistakes are made precisely when rushing to conduct operations. So I needed the final thirty seconds to verify I had not introduced a latent error that would have caused the burn to fail. Rather than double-checking the written procedures, which would have taken too long, I cleared my mind of all extraneous issues and concentrated only on the critical items, based on my technical knowledge of the system and what was required for the burn to succeed:

1. The engines needed propellant, so I verified the switches were in the proper position to allow the pressurized tanks to pass fuel and oxidizer through regulators and feed the combustion chambers.
2. Both engines needed an ignition source, so I verified the ignition system was armed and powered.
3. The thrust vector control system needed proper guidance from the flight control computers, so I verified I had the proper steering data entered into the computers.
4. Finally, the computers needed to know when to send the ignition command, so I verified the time of ignition was correct on the display.

Propellant, ignition, steering, and timing. Nothing else really matters. With sufficient training and preparation, all that matters can be accomplished in fifteen seconds. Any latent errors can be rectified in the few seconds it takes to move a switch or type a command.

In naval aviation, when the operational pressure was off-scale high on an already high-risk mission and I was getting behind in the timeline, I had developed the ability to ignore the small issues and focus on the critical factors to achieve the goal. I used the same technique in the lab that night, and it worked.

I learned two things during that SAIL test, one rational and one emotional. The first was that operations in the Space Shuttle were identical to high-pressure operations around the aircraft carrier. So the mental skill necessary to focus on the critical factors, when the risk is high and errors cannot be tolerated, would be the same, whether on a ship at sea or a ship in space.

The second thing I learned was the pressure I would feel on launch morning, with the cameras rolling and the engines igniting, couldn't be any greater than the pressure I felt to complete the test at 1:00 a.m. in front of twenty-five frustrated, judgmental engineers. It worked then, so it should work on launch morning.

You can't climb onto a rocket without confidence. That night I knew I had developed sufficient confidence in my ability to focus on critical items and eliminate errors as the countdown clock started. I knew I had the proper knowledge, skill, and attitude to strap in. Whether I would execute properly on a rocket after ignition remained to be experienced.

That question of execution in a dangerous environment was the fourth reality that confronted me, but it waited until five months later, on the night before my first launch attempt. (See photo 3, taken the night before a different launch.) This was

Photo 3: *Endeavour*, on the night before launch of STS-113
(Photo Credit: NASA)

the most challenging of my four milestones, and it catalyzed the most valuable and profound change in my mental attitude about risk.

When my head hit the pillow, with the intention of going to sleep, an overwhelming thought entered my mind. I realized *I had run out of time to get any smarter, and it was too late to quit.* Had I prepared enough? Would we die tomorrow? Would I fail and cause our destruction? That would be worse than death. I needed to resolve these questions in my mind to calm down and sleep to perform at peak effectiveness—and stay alive—the next day after strapping on top of 4.5 million pounds of explosive propellant and dangerous systems.

After ten minutes of thinking about the challenges I would face, two thoughts came to me. First, flying in space was the only thing I had wanted to do since I was ten years old. Helpful, but insufficient. I could still die, and that was *not* what I wanted to do since I was ten years old. My second thought was much more useful. If the laws of physics and engineering tried to conspire with the evil gods of death and destruction and something bad started to happen to our spaceship or its equipment, I promised myself I would spend my final seconds trying to save, in priority order, my crew, our vehicle, and the mission.

This was the operational thought I needed to calm down. Nothing else in the world mattered, including me. Save the *crew, vehicle, mission.* That's all I could attempt, and I was ready to do that to the death. And I peacefully fell asleep, armed with a profoundly improved attitude about risk tempered by four significant milestones in a fourteen-year crucible of hazardous operations.

The next morning my crewmembers and I were awakened at a particular minute in the launch schedule with a knock on our doors from Olan Bertrand, Vehicle Integration Test team leader, a man as nice as he is large. From that moment until much later in our flight day, several hours after we were successfully on orbit, I stayed mentally in the present and concentrated intently on what I was doing to make absolutely sure I committed no errors of any consequence.

During the prelaunch flow, I did not concern myself with anything that might happen further than thirty minutes into the future. Be here, now. At a particular time in the prelaunch schedule, I called my wife to say good-bye. As much as I love her, after I hung up I couldn't and didn't think about her.

In the Crew Quarters, we remotely attended the technical systems briefing to hear the final status of the launch and flight systems. This was our opportunity to thank the program managers, engineers, and processing technicians for their care and dedication in preparing the systems for the rigors of spaceflight. I promised the personnel I would try to bring the vehicle back in the same pristine operating condition.

We attended the weather briefing for an update of the current and forecast conditions at the launch site and possible emergency landing sites. I always smiled when we received updates on the forecasted space weather, which included status on predicted solar radiation at our planned altitude in low Earth orbit. No storms on the cosmic horizon.

Some of my favorite support personnel, the cooks who cared for us in Crew Quarters and delivered the meals stowed for us in flight, provided the traditional launch-morning breakfast. Historically, the early astronauts ate substantial meals of steak and eggs before launch. For me, the breakfast was more of a ceremony. Before each of my six flights, I never had much of an appetite, even with the customary cake adorned with the crew patch sitting on the table in front of us.

At a precise minute during the countdown, and for the next three minutes, members of the press were ushered into the wardroom for a photo opportunity. Photo 4 shows a picture of us at the launch-morning breakfast before STS-86. In my case, I can share I am forcing a three-minute professional smile because this is what's required during that period in the launch flow. At every event on launch morning, I stay in the moment, do what is required at the time, and eliminate all extraneous thoughts about the future.

Photo 4: Launch morning breakfast, STS-86 (Photo Credit: NASA)

After breakfast, we climbed into diapers and liquid-cooled undergarments and walked in our long underwear and socks to the suit-up room. Technicians helped us don our pressure suits and conduct final leak checks to verify pressure integrity. These suits were introduced after the *Challenger* accident and were intended to help us survive if the cabin was breached during launch or landing. The orange color of these suits was introduced to help us be spotted by rescue forces if we decided to become bobbing corks in the Atlantic Ocean. The diapers would be needed if the flight systems didn't work as advertised.

At another precise minute during the countdown, the flight crew departed Crew Quarters for the ten-mile ride in the silver Astrovan to the launch pad. (Photo 5 shows this evolution before STS-113.) The long ride can be scary. Though we were with our teammates, we were also alone in our thoughts. Two things helped me. I knew we had received the best training from the best instructors and we were ready. Second, even if our launch time required us to wake up at 9:00 p.m. to start our day, hundreds of Kennedy Space Center workers lined the streets at 2:00 a.m. to cheer us on and wave good-bye as we rode to the pad. Through our tinted windows, they couldn't see us, but we saw them and they always inspired me.

Photo 5: Walk-out from Crew Quarters, STS-113 (Photo Credit: NASA)

I was one of the lucky commanders. On the five flights I commanded, I had to take the long ride to the pad only one extra time when we didn't launch. During each of the six trips, I observed my crewmembers and attempted to gain insight into their mental and emotional state of readiness. In the calm before a dangerous operation, strange humor or uncharacteristic behaviors sometimes surface. I believe this emotional release is a way some people manage fear.

Near the end of the trip to Launch Pad 39-A or 39-B, the Astrovan made one final left-hand turn, and the crew was faced with a scene similar to the one shown in photo 6a. Regardless of how the crewmembers were behaving before the final turn, this view was always sobering and inspiring. An unbelievable feeling occurs when seeing such an impressive site. If the launch was at night, the magnificent engineering marvel that is a spaceship stands on the pad illuminated by twenty-million-candlepower xenon lamps under a beautiful moon.

Every time I experienced this sight, I realized how privileged I was to be selected to support our nation's space program. I also realized at each of those moments every one of my crewmembers was intellectually and emotionally ready to execute the mission with the highest professionalism and operating excellence.

Photo 6a: *Endeavour* on the launch pad, STS-113 (Photo Credit: NASA)

The Astrovan stopped at the base of the launch pad. Again, what helped me to prevent errors and not worry about what I couldn't yet control was staying focused on the present. I can't do anything about the future, which never seems to come until later. As I egressed the Astrovan, the possibility that I might die in a fiery explosion never entered my mind. The only two thoughts I had while climbing down the steps of the van were *make sure my zipper is closed* (sometimes we opened it for extra cooling in the hot suits), and *don't trip and look like a fool.*

While walking to the gantry tower, we saw a view similar to the one shown in photo 6b. *Endeavour* was ready. The crew was ready. Our next ride would be a little longer, a little faster, and a little higher. It was time to leave Earth.

Photo 6b: *Endeavour* on the launch pad, STS-113 (Photo Credit: NASA)

Chapter 3

PRINCIPLES OF OPERATIONS
The Challenges of Launching People into Space

In the decade of the 1990s, the success of NASA in space can be attributed to George W. S. Abbey (shown in photo 7). He led our human spaceflight program and was responsible for restoring operating excellence into the agency after the *Challenger* disaster in 1986. Though he never asked for the honorific, almost everyone referred to him as Mr. Abbey. Those of us who understood the dangers of spaceflight appreciated the leadership skills of Mr. Abbey. We knew he was the reason so many missions were successfully accomplished in Earth orbit, given the technical, political, and social challenges of bringing hundreds of thousands of people together with the common goal of sending humans into space. If not for Mr. Abbey's leadership and technical expertise, any of these challenges would have easily killed the program—and more crews.

In the second half of the 1990s, the administrator of NASA tasked Mr. Abbey to lead the effort to redesign the original Space Station *Freedom*. In that role, he became the father of continual human presence in space, which began on November 2, 2000, with Expedition-One on Station. Mr. Abbey was the single most important influence that resulted in the successful design, fabrication, launch, and construction of the International Space Station.

Mr. Abbey's ascent through the command structure to the top of human spaceflight was launched much earlier as a result of a recommendation given to the president of the United States for a corrective action at NASA. One week after the Space Shuttle *Challenger* came apart during powered flight on January 28, 1986, President Reagan issued an executive order to investigate the accident. William P. Rogers, a former secretary of state under President Nixon, led the investigation commission. The vice chairman was Neil A. Armstrong, former astronaut and commander of *Apollo 11*, the first manned lunar landing mission. The presidential commission was tasked with developing recommendations "to prevent any recurrence of the failure related to this accident," and "to the extent possible to reduce other risks in future flights."[14]

Photo 7: Official NASA photograph of Mr. George W. S. Abbey (Photo Credit: NASA)

The commission issued its report on June 6, 1986, with nine recommendations for the president to consider. Armstrong, better than anyone on the planet, understood the operating excellence required to lead the complex sociotechnical effort to explore space. During its investigation, the commission members recognized that, as they wrote in their report, "…there appears to be a departure from the philosophy of the 1960s and 1970s relating to the use of astronauts in management positions. These individuals brought to their positions flight experience and a keen appreciation of operations and flight safety."[15]

At the time of the accident, Mr. Abbey was the director, Flight Crew Operations, but he and his astronauts had less influence in the decision-making process within the management structure of the Space Shuttle program than they had during the *Apollo* era.

About one month after the investigation report was published, two rookie astronauts, Mike McCulley and Bill Shepherd, performed a simultaneously symbolic,

14 Report, Presidential Commission on the Space Shuttle *Challenger* Accident, June 6, 1986 (Washington, DC).
15 Ibid.

sarcastic, and humorous act to indicate that some NASA managers thought we were suddenly in full compliance with one of the commission's nine recommendations. At 6:00 a.m., McCulley and Shep smuggled four concrete cinder blocks into Mr. Abbey's office in the main building of the Johnson Space Center, the home of human spaceflight for NASA. They hoisted his desk on top of the dusty concrete blocks and placed a placard on its front for all to read:

> In compliance with Rogers Commission Recommendation II:
> "The function of the Flight Crew Operations director
> has been elevated in the NASA organization structure."

Mr. Abbey kept his desk in this stand-up configuration, similar to Winston Churchill's, as a personal reminder of his high responsibility to restore operating excellence to space exploration at NASA. Gradually over time, Mr. Abbey's voice was heard, and he was able to use his leadership skills to influence a resurgence of technical and managerial excellence into human spaceflight operations.

The Hazards of the Job

Almost a decade later, at the height of operating excellence in the Space Shuttle program under Mr. Abbey, I had a conversation on orbit with astronaut Mike Foale about the difficulty of sending people into space. Mike and I are shown together in photo 8, taken during STS-86. Mike was onboard the Russian Space Station, *Mir*, when it suffered a near-catastrophic collision with a *Progress* supply vehicle, which caused a breach in the hull. Mike and the two cosmonauts had only ten minutes of

Photo 8: Mike Foale (on the right) with the author, STS-86
(Photo Credit: NASA)

air remaining when they successfully isolated the leak from the *Spektr* module, which was attached to *Mir*.

What triggered our thoughts about the challenges of sending people into space was an opportunity Mike and I had to observe a rocket on its ascent trajectory launching from Earth. It's hard to beat the experience of seeing a launch in person. Surely, though, one better way is to experience the view of a launch in your rearview mirror, while strapped onto 4.5 million pounds of hardware and dangerous propellant.

In October 1997, Mike and I had the most rare of opportunities. We had the double privilege of already being in space when we saw another rocket launch from Earth coming up into our orbit. As we experienced the reality of seeing something we had only imagined in dreams or science fiction, we began to discuss the technical engineering complexity, the immense energy required, and all the mechanical and human elements that have to work properly to launch equipment and people into space.

First, the kinetic energy requirements. The winged orbiter alone weighs 240,000 pounds, or 120 tons. It needs to be accelerated to a speed of 17,212 mph, or 25,244 feet per second, to achieve orbital velocity (STS-113 data). To accomplish this, we use five rocket motors.

Three Space Shuttle main engines, the most efficient aerospace engines ever created, burn propellant at a rate of 1,000 gallons per second through nozzles operating at 6,000°F. To keep the nozzles from melting, the nearly frozen propellant is sent through tubes that line the nozzles. In a double benefit, some of the propellant becomes preheated in this circulation process.

The two solid rocket boosters are monsters, and they are scary bad. The solid propellant weighs one million pounds in each booster, measuring 149 feet long and 12 feet in diameter. Aluminum powder is used as the fuel and ammonium perchlorate as the oxidizer. To adhere these two materials, a bonding agent is used called polybutadiene-acrylic acid-acrylonitrile. The propellant, which assumes the look and feel of rubber, is shaped into an eleven-pointed star. The spaces between the points, along the entire internal length of the white rockets, form the combustion chamber of Hades. Each booster generates three million pounds of thrust. Once they are ignited, they burn to depletion, with no questions asked and no secondary plan.

Two weeks before liftoff, the Space Shuttle starts its journey at a near sedentary one mile per hour as it is transported from the Vertical Assembly Building to the launch pad on a crawler, known as the Mobile Launch Platform. (See photo 9.)

On this slow ride to the pad, a future commander and crew can stand back by the tail of the orbiter and see the tip of the vertical stabilizer swaying back and forth

Photo 9: The Space Shuttle *Discovery* on Mobile Launch Platform-1 before STS-102 (Photo Credit: NASA)

about three inches because of the bumps in the road. That crew can wonder how such a fragile vehicle is going to keep together during the ferocious launch event a few weeks later. They won't come up with a satisfactory answer.

At the end of the six-hour journey, to position the stack precisely on the launch pad, oriented with its belly toward true north, spin-off technology from the farming industry is used in reverse. In this case, instead of sharing NASA-developed high technology with other industries, we borrowed the advanced technology from farmers. An optical system that is used on plows in the fields of Kansas, to keep the rows of corn and wheat laser straight, positions the giant stack within an inch of where it needs to be for liftoff.

Liquid oxygen and liquid hydrogen are stored in separate tanks, for safety reasons, on opposite sides of the pad. The night before launch, these two liquids, explosive when combined, are brought together carefully and stored in the orange external tank at minus 423°F. As Dan Brandenstein, the commander on my first mission, explained, the launch team will then clear the pad of all nonessential personnel except, "You seven, come on up here."

Photo 10 shows the final walkway across the gantry leading to the white room for ingress to the vehicle. If you are one of the seven and you look to your left just before entering the white room, you'll see a single word, printed on the side of the solid rocket booster in two-inch-high, block letters, that reads simply, "LOADED" (see photo 11). For a moment, this word makes breathing difficult. School's out; it's time to go to work.

Photo 10: Gantry access to white room 195 feet above the launch pad

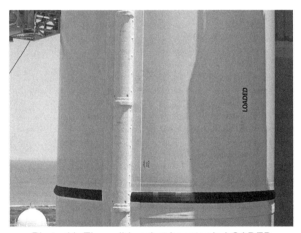

Photo 11: The solid rocket booster is LOADED

The entire stack sways in the wind as the crew crawls in; the assembly is held vertically only by eight bolts and nuts, 120 feet below at the bottom of the solid rocket boosters. The bolts are about four inches in diameter, and each frangible nut weighs more than sixteen pounds.

Three minutes and twenty-five seconds prior to liftoff, the main engine gimbals are swiveled by hydraulic actuators commanded by the flight control computers, and the vehicle starts to shake. Fourteen seconds before liftoff, millions of gallons of water are dumped into the flame trench. The crew feels the waterfall, with vibrations coming up the stack. This water is intended to take the brunt of the explosive start and will dissipate the acoustic and thermal energy after being vaporized and blasted outward in the giant steam cloud visible for miles. If not for the water, now steam, the acoustic energy and vibrations would reflect back as shock waves and tear the vehicle apart before it ever rises above the tower.

Six and a half seconds before liftoff, the three main engines are started. They spool up to 90 percent power in three seconds, and computers check thousands of parameters to verify the engines and vehicle are flight worthy. If they are, the engines are throttled up to 100 percent. The whole stack sways forward about two feet at the top, colloquially called the "twang," under the bending moment from 1.5 million pounds of thrust, still harnessed.

If the computers are satisfied, when the clock hits zero, the fuse is lit. NASA standard detonators fire up charges along the internal surface of the solid rocket fuel. Internal pressure explodes to 900 pounds per square inch (sea level air pressure is 14.7 psi), and the internal temperature reaches 6,000°F (surface of the sun is 10,000°F).

Liftoff is a violent event that will occur after the solids are ignited, even if the frangible nuts on the hold-down posts fail to disengage. A thrust of 7.7 million pounds will rip through the unbroken nuts, regardless. Launch and ascent is the most sustained, straight-lined power a human can experience in an uncontained environment for more than eight minutes and come out alive at the other end.

Other than emergency personnel, the closest humans are three miles away. From that distance, liftoff looks slow and stately, and the vehicle appears to climb methodically. The distance is deceiving—there is nothing slow about a launch. Instantly, the astronauts are slammed into their seats with twice the force of gravity. The launch tower, visible in their peripheral vision, disappears. The feeling is analogous to lying down on a hard bench in an elevator on the nineteenth floor of an office building, 195 feet above street level, and having the floor suddenly open up into the elevator shaft. The speed and acceleration are identical, except everything is going straight up, faster and faster, instead of down.

I was told once you could tell how powerful an afterburner was by the length of the plume compared with the length of the airplane. The fire coming from the solid rocket boosters is about three times the length of the Space Shuttle. So the stack can generate three times the force of gravity, much more than atmospheric afterburners. By the way, all that fire is being converted to kinetic energy, which must be removed at the end of the mission if you want to come home. The only way to do that is to reenter the earth's atmosphere and slam into the molecules of oxygen, rip them apart, and use friction to slow down, which creates a twenty-five-mile-long trailing plume of fire. What goes up must come down, with equal energy.

At the moment of launch, I weigh 195 pounds in my liquid-cooled underwear, but I am also wearing an 85-pound pressure suit. Add it up, multiply by two, and the result shows I weigh 560 pounds strapped into the seat. As powerful as the rocket is during launch, it is actually being patient for the first forty-five seconds of ascent. If it pushes too hard, the wings and tail of the orbiter rip off. Not good, if you want to land on a runway two weeks later. So the rocket throttles itself back, biding its time, until above the thick part of the atmosphere. This occurs less than a minute after liftoff.

Now the astronauts are in for the ride of their lives. The main engines spool up to 104 percent. The internal shape of the eleven-pointed star in each booster exposes more surface area of burning propellant, which suddenly allows the boosters to push harder and harder. At two minutes into flight, near the end of the first stage, the boosters are almost empty. The reduced weight allows them to push with three times the force of gravity. Now I weigh 840 pounds. And it feels like it.

With personal oxygen bottles hanging off the side of the seat and pulling on my chest harness, it feels hard to breathe. The trick is to never allow my lungs to deflate, or I won't get them re-inflated. I must use short, panting breaths with my diaphragm.

If the vehicle deviates from the desired trajectory, a nice man called a range safety officer, who sits at a console in a military bunker, sends a destruct signal to explode the stack over the ocean. After all, we don't want to injure any of the US population because of a mistake we made. A few months before launch, my wife directed me to meet with Officer Red Button and show him pictures of our children (see photo 12) to help him make a good decision on launch day.

At two minutes and ten seconds, the boosters are spent. I make the mistake of looking out the side window to see them separate and am shocked by the sight. With a loud explosion, two smaller rockets mounted on the bigger rockets blast toward me to push the bigger ones away from the orbiter. As the boosters fall away, I decide it is better to focus on the instruments in the cockpit and anticipate any of the hundreds of smaller failures that might try to escalate into one larger failure.

The three main engines must be able to function in space where there is no oxygen, so we must carry our oxidizer with us in the orange external tank, rather than using the oxygen in the atmosphere on our way uphill. The most efficient way, using modern technology, to accelerate the stack fast enough to

Photo 12: The wallet-sized photo of my children, intended to influence the range safety officer to make the right decision

achieve orbital velocity while carrying all that weight is to go straight up and get out of the thick part of the atmosphere as expeditiously as possible. As soon as we don't need to worry about atmospheric drag, we tip over and go as fast as we can horizontally. Photo 13 shows the ascent trajectory in a time-lapsed photograph.

The vehicle is climbing the whole time in the picture. The path only looks like it is descending because the vehicle is so far downrange that the earth has begun to drop away from the local horizontal where the camera is located.

On we go, faster and faster, harder and harder on our ascent trajectory. At about nineteen times the speed of sound, I see a piece of ice outside my forward window that has been knocked from the cold external tank under vibration and is accelerating *forward*, faster than three g's. For a quick second, the sight doesn't make sense. How could a piece of ice, without engines, be accelerating faster than we are, with our enormous engines? But then, I realize we are in the vacuum of space, and the ice is bodysurfing forward on our exhaust wave from behind.

Photo 13: Ascent trajectory of STS-113 (Photo Credit: NASA)

Here's how the phenomenon occurs. Seven minutes earlier, as we climbed from thick atmosphere at the surface of the earth to thin during ascent, the exhaust plume behind us was widening. By the time we were in a vacuum a few minutes later, the plume had widened so far that it came all the way around and surrounded us. Exhaust from our engines is now pushing out, away from our vehicle in *all* directions, including forward, in the form of an external pressure wave. In the one and a half seconds I spend looking outside, the thought occurs to me this big-wave-surfing ice is visually demonstrating how thrust works in a vacuum.

The powerful exhaust plumes from our engines can't push against anything behind us to make us go forward—nothing is back there. But that's the point. The pressure from the burning propellant *inside* the combustion chambers of our three main engines is pushing hard against the internal forward wall of each chamber at its front end, and there is no internal wall to push against at the back end because that's where the open nozzle is conveniently located. Thrust is actually this huge unbalanced push on the forward walls inside the combustion chambers, which is in turn pushing forward on anything strapped onto the chambers with the force of 1.5 million pounds, enough to achieve orbital velocity (while making it hard to breathe, but that's a secondary concern).

One hundred and twenty tons of hardware, software, and jellyware—we are there to make sure the hardware and software perform properly—are accelerated to twenty-five times the speed of sound. It is difficult to believe you can get that much mass moving that fast. By the time we run out of propellant, after eight minutes and twenty-three seconds of powered flight, I am surprised we aren't halfway to Mars. At this point, we are instantly weightless. On my first flight, I could tell we were weightless because a piece of a cookie from the previous crew, and a stray bolt from something somewhere, floated up in front of my face. I only hoped the bolt wasn't the one that was supposed to hold the whole orbiter together.

On my fourth flight, I arrived on orbit four months after Mike, though we experienced the power of liftoff and ascent in the same vehicle, *Atlantis*, on successive launches. We found it hard to believe we survived our separate trips, were floating together in orbit, and were watching yet another rocket launch from Earth toward us in space. For about twenty minutes, we had a sobering conversation about the thousands of small and large engineering miracles that had to occur in a precise sequence to launch things and people into space. Mike eventually concluded, "It's best not to think about it." So, we quietly went back to work in our little office cubicle, humming around the earth at five miles every second.

Predicting Failure

In the late 1990s, Mr. Abbey intuitively sensed the Astronaut Office was beginning to drift toward a potential accident. His intuition was founded on his extensive experience as an engineer and manager in the *Apollo* program, with stellar mentoring he received from Dr. George Low, who ran the program. Dr. Low was responsible for restoring operating excellence in NASA after the fatal fire that killed Gus Grissom, Ed White, and Roger Chafee during a launch pad test in AS-204 (later designated *Apollo 1*).

Legendary stories of Dr. Low's leadership were still told thirty years after he made the courageous decision in 1968 to send three men away from our home planet's gravitational influence for the first time in its 4.5-billion-year history. *Apollo 8* was a demonstration flight to test and verify all operational aspects of the *Apollo 11* lunar mission except the final landing on the moon's surface. Mr. Abbey learned from the best, as he assisted Dr. Low, and experienced the most effective ways to ask insightful questions and provide leadership to a collection of the finest engineers and managers in this world on a mission to send operators to a different world.

As with all great leaders, Mr. Abbey developed his leadership skills from working with and learning from multiple leaders and mentors. Dr. Robert R. Gilruth, and the legendary Christopher Columbus "Chris" Kraft Jr., giants in the early space program, were particularly influential in shaping the leadership style of Mr. Abbey. He cites Major General Yuri Glazkov, Russian Air Force, the head of Russian spaceflight crew operations, as another of his mentors. Mr. Abbey humbly admits that any success he had came from practicing what he was fortunate enough to have learned from these individuals and from his association with two truly great engineers, Dr. Maxime A. Faget and Captain John W. Young, US Navy (Ret.).

Dr. Faget designed the shape of the *Mercury* capsule in his head and documented it initially on the back of a bar napkin, or so the legend has been passed down to us in the later generation. Capt. Young is the most brilliant, intuitive, and insightful engineer and test pilot I have ever known. When anyone introduces John, the ultimate professional, as an astronaut who walked on the moon, he always responds, "No, I *worked* on the moon."

Capt. Young is incredibly funny, yet he never understood why so many people think that. In the mid-1980s, I was riding in a car with John and astronaut Lacy Veach, who were both reading newspapers. Indicating he thought the Russians likely stole the design of our Space Shuttle, Lacy suddenly announced sarcastically, "Hey, John, look at this; the Russians have a new Space Shuttle that looks just like ours!" John was the consummate test pilot and nothing ever fazed him. As the commander of the first flight of the American Space Shuttle, he had intimate knowledge of its many design flaws. Without even looking up from his paper, John said simply, "Serves them right."

The best leaders, such as Mr. Abbey and Dr. Low, have developed the skill of flipping hindsight into foresight to predict and prevent potential accidents. Although he created and inspired so much success in the human spaceflight programs, Mr. Abbey was never satisfied in his pursuit of operating excellence. He began to detect subtle signals that indicated the Astronaut Office was slowly drifting toward its next potential tragedy, specifically while astronauts were conducting spaceflight readiness training in the NASA T-38 aircraft. The astronauts knew well the dangers of spaceflight. They continued to respect the complexities of the Space Shuttle systems, as evidenced by their professional discipline and rigor in preflight training and preparation for space missions. But some astronauts had begun to operate in a way that indicated indifference or complacency when operating the much simpler, yet equally deadly, T-38 airplane and systems.

In 1998, Mr. Abbey asked me to be the director, Flight Crew Operations. This was the management position first held by Deke Slayton, one of the original *Mercury 7* astronauts. I felt I was not ready for the job. Mr. Abbey thought I was. He gave me one command before I took the controls: Don't let them have an accident.

In that one statement, Mr. Abbey taught me a valuable lesson. He disagreed with some of the other executives in the space program, who were driving aggressiveness when they said: "Our primary mission is to 'fly safely.' But, remember the first word is '*fly*'; people are depending on us to support the flight schedule." Those aggressive statements influenced middle-level managers to take risk unnecessarily. Similarly, other executives had invoked St. Thomas Aquinas, "If the highest aim of a captain were to preserve his ship, he would keep it in port forever." In both cases, the intended message from those other executives came through loud and clear to employees and contractors: Accept the risk, fly, complete the mission, do whatever it takes, get it done. The dangerous characteristics of the can-do culture were regenerating ten years after the *Challenger* disaster.

The way those executives spoke, intending to motivate employees to accept risk and aggressively pursue the mission, is exactly how most bosses talk to their employees today in many organizations, especially with increasing competition and pressure to deliver results. Such organizations are headed for trouble. Certainly, the primary mission *is* paramount. Executives and managers give commands to their teams in terms of the organizational *results* they want. In the case of spaceflight, the executives wanted the middle-level managers to "support the flight schedule." Just get it done.

The better leaders try to inspire their followers by attending to the *quality of the process*. Quality is lost when mediocre bosses speak extensively about results. Even if these bosses add a few cursory and compulsory comments about safety, as in, "Remember, do it safely," the workers know what the bosses really want. Contrary to what the bosses say, the workers believe the main message in those speeches. Those bosses are only interested in results—accept the risk; *get it done!*—and are not interested in the quality of the process.

Elite leaders talk to their workers differently. We can all learn from watching how the best leaders lead. In Mr. Abbey's mind, the primary mission was still paramount, but he changed the definition of the mission. He strongly believed the primary mission was not solely accomplishing flights. The primary mission was performing flights *correctly*, with operating excellence, to achieve the best possible results over the long term. Yes, we do need to explore space. It is our human destiny. But, we don't need to explore space on Tuesday; we can explore space on Saturday after we have fixed our problems, and we are truly ready to fly.

Mr. Abbey taught me another lesson worth mentioning. In the three years I held the position of director, I was required to meet with him every two weeks to provide my assessment of the operational readiness of the Astronaut Office. When I reported examples of success or high-quality operations, he often appeared uninterested or even bored. In the times when I reported issues, problems, or challenges, his face lit up with engagement. Even if I didn't have a proposed solution, or suggestion on how I might address the issue, he expressed genuine satisfaction that at least I had identified an issue that could be fixed.

This may be one of the more powerful techniques a leader can use in a dangerous business. My leader gave me positive reinforcement to *search for vulnerabilities* in my organization. Too often, bosses give positive reinforcement when their subordinate managers provide "sunshine reports" or glowing statements of success. The only value in a sunshine report is in knowing the subordinate manager is not looking hard enough to identify vulnerabilities. In a business, failing to identify issues will result in lost productivity. In a dangerous business, failing to identify issues will kill people.

Changes in the Organizational Culture of the Astronaut Office

As the new director, Flight Crew Operations, in September 1998, my responsibility was to improve the operational readiness of the Astronaut Office. By that time, I had been flying with NASA for fourteen years and was privileged to have flown four times on the Space Shuttle. So I had developed an opinion of the challenges we faced in the office's operational readiness. But I didn't want to develop corrective actions to improve the way the astronauts were operating if I was using erroneous assumptions or incorrect analyses to derive faulty conclusions.

Through the years, I watched good leaders make their best decisions when they solicited, encouraged, and valued the opinions of others. I convened a small team of astronauts to assess the readiness issues in the astronaut corps related to training and flying on the Space Shuttle and in the T-38 airplane and to make recommendations for corrective actions to reestablish a culture of operating excellence. In our first meeting, we discussed the current state of operational readiness in the Astronaut Office.

The assessment of the current culture in any organization should begin with a good understanding of how the organization became successful when it was formed. In 1959, NASA selected the original *Mercury 7* astronauts. A small group of NASA managers chose from the best candidates who were already military test pilots, for one important reason. Given the dangerous mission and the accelerated schedule, with little time available for basic flight training, the managers on this first selection board decided to select pilots who had already demonstrated they innately understood the principles of operating excellence in hazardous flight environments.

Over the next thirty-eight years, through the Gemini, Apollo, and Space Shuttle programs, seventeen more groups of astronauts were chosen. The leaders in the Astronaut Office did not feel the need to describe verbally, or document in writing, the principles of operating excellence in flight operations. The values and beliefs that made astronauts successful were tacitly passed along through the cultural norms of the office from one group of astronauts to the next.

As our first meeting continued, the team discussed how the operating culture in the Astronaut Office had degraded gradually. I was pushing the team to jump straight into describing the best techniques we wanted astronauts to employ in flight operations to prevent accidents and return to operating excellence in training and in flight. At that point in our team's meeting, Ken Cockrell, an experienced test pilot and astronaut, made an insightful observation with a recommendation. We had never captured in writing an agreed set of principles of operating excellence. We should not attempt to describe the best techniques to be used in flight without first describing the underlying basic principles of operations for spaceflight crews.

Here is an explanation of how the culture in the Astronaut Office was changing. During the Space Shuttle era, the emphasis on science was increasing. More astronauts were selected from the best candidates in the fields of engineering, science, and medicine and fewer from military aviation. The candidates were outstanding in their respective fields, and they came with their individual value systems based on their personal successes in their fields of expertise. One of the more powerful embedding mechanisms for the shared values in a culture is the selection process. Over the years, smaller percentages of candidates were chosen from the ranks of military test pilots. Silently, the original values and beliefs of operating excellence in hazardous flight environments had begun to seem less important for creating success with the new generation of astronauts, who were exceptional engineers, scientists, and doctors.

Cockrell suggested, if we were going to make changes, the first should be to codify the principles of spaceflight operations previously unwritten but innately understood by the early astronauts. I agreed, as I didn't think the original principles were understood or were being followed by some in the newer generation of astronauts. As leaders, we had failed to reinforce the importance of the original values and beliefs that created success in hazardous aviation environments. Only from the basic *Principles of Operations* for spaceflight crews should we begin to develop the *Techniques for Operating Excellence*.

The Fourteen Principles of Operations for Spaceflight Crews

About halfway through our first team meeting, we suspended our collective work of discussing the current culture and developing techniques so I could work alone and describe the high-level Principles of Operations for spaceflight crews that I wanted the astronauts to understand and follow. After reestablishing the desired original principles, we could continue our meaningful dialogue, over a much longer time period, to develop the Techniques for Operating Excellence to be used by spaceflight crews, during ground training and in flight, to prevent accidents and improve performance.

The following are the Principles of Operations for spaceflight crews I wrote (including one that astronaut Charlie Precourt wrote) that day. These principles were based on the elite training and mentoring I was privileged to receive during twenty-three years, while operating and flying with the world's best leaders and aviators on the ground, at sea, in the air, and in space, with the US Navy and NASA.

1. The commander is responsible for the development of optimum crew resource management.
All crews are different, and each crewmember possesses different skills. The commander is expected to be able to identify the operational techniques and methods used that achieve the best possible human operator performance.

2. The commander must use good leadership skills.
A good leader brings out the best in the crew. He or she sets a good example of how to operate. The leader has superior judgment, unquestioned honesty, and common sense. Although the leader will involve the crewmembers in the decision-making process, he or she will be decisive when required and will make good operational decisions. A good leader always ensures the crewmembers clearly understand the various mission objectives and that individual responsibilities are clearly defined. The leader always attempts to be fair, is disciplined, and enforces discipline in the crew.

3. All crewmembers must use good operational techniques to enhance mission success.
Crewmembers must use the keyboards, switches, and controllers effectively. They know what to memorize or think about during critical periods. They know how best to control the vehicle, robotics, payloads, and equipment. Crewmembers share information and techniques they have learned with their crew and discuss the advantages and disadvantages of using these techniques. They follow and effectively execute the flight data file (FDF) procedures, and they do not change the procedures without concurrence from the Mission Control Center.

4. All crewmembers must operate as part of a team.
To get the most effectiveness from a team, crewmembers must share knowledge and help each other. Crewmembers must not highlight their own performance at the expense of another. We must get the most from every member of the crew. Each

crewmember should have responsibilities that challenge his or her abilities and encourage coordination and teamwork. More experienced crewmembers should consciously set an example of good teamwork to newer flyers. All our operations are based on operating as part of a team—respect for each other and for our training and Mission Control Center teams is a necessity.

5. Crewmembers must understand the flight and payload systems.
We have a responsibility to fully understand the assigned flight and payload systems. This knowledge will save lives and will enable mission success when things go wrong or if unplanned situations occur. Knowledge of systems will help to identify and avoid errors if incorrect directions are received from Mission Control Center, if the FDF is in error, or if the crewmember begins to make a mistake.

6. The crew must foster team support from the Mission Control Center.
The Mission Control Center is a great resource. There are many talented people in the Control Center, and in some cases, they have more data than the crew on board. They have much experience and very detailed knowledge of their individual systems, with historical relevance of previous flight operations. Even though the operators in the Control Center do not directly experience the flight environment, we must make every effort to fully use their expertise and reinforce their importance to the team.

7. Crewmembers should communicate effectively with the Mission Control Center.
Communication with the ground personnel must be clear and precise, with no ambiguity. Misinterpretations must be resolved immediately, and there should be no doubt or uncertainty in verbal transmissions. Also, remember with an international space program, the possibility of errors is greater since there will be multiple languages being spoken by crewmembers with differing levels of communication ability.

8. The crew must solicit constructive criticism from the instructor team.
The instructor team members give us valuable training. We are responsible for drawing from this valuable resource. Arguing over details should be saved for polite, off-line discussions. The example set by the commander here is critical, particularly for newer crewmembers. It also sets the tone for the instructors. Feedback ends up being based on the crew's reaction. *Defensive* crews do not get the same quality feedback that *open-to-suggestion* crews receive.

9. Crewmembers should ensure methods exist for identification and rectification of errors or omissions.

There are many good techniques to accomplish the task of error management. One of the best ways to minimize the consequences of errors is to use the two-person rule where both crewmembers work together reading the FDF procedures and verifying proper actions. This is often a misunderstood and misused concept; if one crewmember simply reads the procedures and the other follows the instructions blindly to move the switches, neither crewmember is providing backup verification of the proper FDF procedure steps.

Good techniques must be used when making switch throws or control inputs, which allow other crewmembers to *trap* or correct errors before they result in any consequence. An example of this is to "telegraph" intended motion by momentarily placing fingers or an open palm on the opposite side of a switch or controller before deflecting in the desired direction. This allows the other crewmembers to see if you are about to move the switch or controller in the wrong direction. We must know how to use the procedures and checklists properly and efficiently. Other useful techniques include using on-board timers and keeping the Mission Control Center involved in on-board operations.

10. The crew should have good methods for handling dissenting opinions.

This is a very difficult problem to solve and deal with well. Throughout the history of aviation, there are many examples of accidents caused by crewmembers not being able to handle and resolve dissenting opinions. This is a critical skill that, if employed properly, may save the mission, the vehicle, and the crew. The commander must encourage all crewmembers, no matter what experience level, to speak up when necessary.

11. At least one crewmember should have situational awareness at all times.

Again, this is a critical concept that should be mastered. Our system has very complicated flight profiles that are not easily understood, nor easily monitored. The on-board displays are not optimally designed to allow situational awareness. This is one of the reasons we fly in the simulator so much before flight. Conducting crew tag-ups is a good technique to keep other crewmembers aware of the operational situation.

12. The crew should have good and efficient techniques for recovering from breakdowns in operational management and situational awareness.

Failure scenarios can be very complicated, and we can easily experience breakdowns or loss of situational awareness. We must be able to quickly and efficiently recover complete awareness of the flight situation. This enables us to be ready to take corrective action if the flight systems fail.

13. All crewmembers should have methods to recognize fatigue and other stresses in themselves and their crewmates and have a plan to cope appropriately with these problems during operations.

The problem of fatigue can be insidious. The onset of fatigue can be difficult to identify unless all crewmembers know themselves and each other well and are constantly making an assessment. They must always be prepared to minimize the effects of fatigue or stress by using good techniques.

14. The crew should have an execution plan for all critical or complicated operations during flight.

During spaceflight, time criticality of some operations is important. Whatever the length of mission, we must be prepared for all critical events. On the long duration flights, we must prepare for critical operations before commencing and then execute the plan well. On the short duration transportation flights, we must not rely on the ability to figure out the best way to accomplish a task on orbit. We must have a plan before we launch, and then we must execute that plan.

The Flight Assignment Process in the Astronaut Office

My next goal was to influence and inspire the astronauts to follow and embrace the Principles of Operations for spaceflight crews. To accomplish that objective, I needed to use the primary mechanism for motivating an astronaut: the flight assignment process.

But first, I had to address some perceived issues with the assignment process. From the time I joined the astronaut corps in 1984, I thought some astronauts were more concerned with the politics of flight assignments, rather than what they could control. That is, some astronauts spent far too much time talking about what the managers were looking for and how to position themselves politically to receive an assignment for a specific flight. Although I understood why they were concerned about their first flight assignment, which is the most important reward in the profession, I was more concerned with something else when I was a new astronaut.

As confident as I was in my abilities as a pilot, I seemed to be more afraid I would not perform well if I were assigned. This apparent conflict was a powerful motivator, which originated that one dark night behind the ship when I allowed my high confidence in night operations at the aircraft carrier to outrun my capabilities, and I lost concentration, resulting in a hard landing. This internal contradiction or cognitive dissonance, where I simultaneously felt confident in my overall capabilities yet lacked confidence in my potential performance in any specific situation, caused me to worry not at all about a flight assignment. Rather, the contradiction caused me to keep my head down studying in the books and practicing in the simulator as much as the NASA system (and my family) allowed. I simply assumed that if I became the best astronaut in the simulator, they would have to assign me to a flight. But above all, by the time I launched into space, I had better be ready to perform well to justify the privilege. Or die trying.

The chain of command that made the flight assignments was known. How the chain made the assignments was unknown. Many astronauts speculated. The sequence of who made the decisions was simple. The chief, Astronaut Office, made the *nomination* for an assignment to the director, Flight Crew Operations. The director made the *recommendation* to the center director, Johnson Space Center, the lead center for human spaceflight in NASA. The center director *approved* the recommended assignment, or disapproved. Although there were some official conversations higher up in the NASA chain of command, as far as the astronauts were concerned, the effective ceiling was the center director. If he approved, you would eventually be strapped onto seven million pounds of thrust. If he disapproved, you would not. So, in the business of flying in space, George Abbey held the keys to ignition.

How the decisions were made was less straightforward than who made what decisions. The chief, Astronaut Office, assessed the skills of the individual astronauts. When I became the director, I was allowed to select the chief who reported to me. I chose Colonel Charles J. Precourt, US Air Force (see photo 14), who is one of the finest leaders I have ever known. The chief reviewed the requirements of the missions some years before each flight. He tried to combine individuals with the necessary complementary capabilities to make a strong flight crew team and match the team to the requirements of the mission. If the mission was highly scientific, the flight

Photo 14: Charlie Precourt on-board STS-91 (Photo Credit: NASA)

crew needed to have strengths in science. If the mission was an assembly flight, with construction requirements to build the International Space Station, the crew needed to have strong capabilities in conducting space walks (extravehicular activities).

As director, Flight Crew Operations, I reported to Mr. Abbey, the center director, Johnson Space Center. My responsibility was to make recommendations to the center director regarding flight assignments of individual astronauts for specific missions. Charlie and I were required to provide rationale for each assignment we were recommending. There was an incorrect assumption in the office that Mr. Abbey micromanaged the assignment process and selected his favorite personalities to fly. In the three years Charlie and I submitted recommendations with rationale to Mr. Abbey, not one flight assignment was ever disapproved. I know he put us in the roles because he trusted us to make appropriate recommendations that matched the right people with the proper skills for each mission.

Strategy to Improve Operations in the Astronaut Office

Col. Precourt and I developed a strategy to improve performance and manage risk in the Astronaut Office. Our goal was to inspire every crew to perform the missions with operating excellence by following our Principles of Operations for spaceflight crews.

The Space Shuttle Program Office was responsible for the quality of the technical systems. In the Astronaut Office, Charlie and I were responsible for how the flight crew used the technical systems. Our two-sided strategy involved *organizational* improvements our astronaut managers would make and *personal* improvements we would ask individual astronauts to make.

The two elements of our strategy, organizational and personal, were not developed or implemented separately. Rather, the organizational and personal sides supported each other. When we set personal expectations for astronauts (for example, to develop interpersonal skills and exhibit the proper attitude), we simultaneously developed and provided an organizational system, in the form of a training process, to help the astronauts fulfill those expectations. And when we developed an organizational process (for example, a qualification process for commanders), we simultaneously considered and tapped into the motivations of individual astronauts, by assigning them to flights when they used the organizational processes correctly.

As director, my responsibility was to set the expectations for all astronauts to follow. When implementing a strategy and attempting to change the way operations are conducted, good leaders publish the expectations for all personnel, so each person knows what to do. Clear terminology is used to allow the people in the organization to develop a new lexicon and have productive dialogue about the progress of change in the organization. Eventually, the personnel form new assumptions about how to be successful in the organization. When the group collectively adopts these new assumptions and beliefs that lead to their new success, the culture has changed. The new culture can be a force multiplier to guide the behaviors of the personnel, even when leaders are not present, to maintain excellence in operations.

One of the common complaints heard around the office was that astronauts did not know what criteria the managers were using to select individuals for flight assignments. The greatest source of motivation for astronauts was assignment to a spaceflight. So I decided to publish the expectations for astronauts in terms of the criteria we managers would use to make the flight assignments. With expectations documented, Charlie and I could use consistent terminology when we attempted to motivate the astronauts to exhibit the desired behaviors to conduct excellence in operations, while preventing accidents. We published the expectations for astronauts in a policy note, "Astronaut Office Conduct and Performance," on August 18, 2000 (included in appendix D, for reference). In the Astronaut Office, we already had formal programs for teaching and assessing knowledge and skills. This note was intended to address the misperceptions in the flight assignment process and to initiate our effort to shape and influence the mental attitudes of astronauts. Attitudes represent the most important element in the triad of knowledge, skills, and attitudes necessary for restoring excellence in operations.

Building Operating Excellence in the Astronaut Office

Process Improvements for the Organization

From the preceding fourteen Principles of Operations for spaceflight crews, Chief Astronaut Charlie Precourt led the development and implementation of several new programs and processes to infuse operating excellence into the Astronaut Office. These new processes, listed below, represented the organizational systems that Charlie and I were responsible for providing to the astronauts collectively to help them be successful. In each program, for the first time in the forty-year history of the Astronaut Office, we published the requirements for astronauts to become qualified and therefore eligible for selection into the specified role.

- Commander's Upgrade Program
- Instructor Astronaut Program
- Skills Development Programs for:
    o Extravehicular Activities Training
    o Robotics Training
    o Rendezvous Training
    o Shuttle Systems Currency Training

After major disasters or minor accidents, organizations typically default to developing new and improved organizational *programs* and technical *processes*. But for us, the development of these programs and processes was only part of our story of change needed in pursuit of operating excellence in our dangerous operations. The other improvements we needed were the *personal changes* in the way *individual* astronauts were using the programs and processes in the complex sociotechnical system. Organizational programs and technical processes are usually easier to develop and implement than the personal changes individuals need to make.

Operating Improvements for Individuals

The improvements in the operating techniques used by individual astronauts were more difficult to address and inspire than the processes. Many books have been written about the organizational or technical process changes that can be used to prevent accidents. Less has been written about how to help individual operators to make the changes needed to improve their own personal performance and prevent accidents in hazardous environments.

In the remainder of this book, I will present the improvements in operating techniques for individuals necessary for controlling risk in our pursuit of operating excellence. Astronauts collectively developed some of the following operating techniques in five decades of operating excellence performed since 1961 in space and on the moon, while facing extreme and immediate danger. Over the years, I included other operating techniques I learned from great operators in the US Navy and in the oil-and-gas industry.

I don't believe these personal *Techniques for Operating Excellence* have been published or explained as a collection. I am certain these techniques have saved my life and improved my operating performance during thirty-five years in hazardous operations. I believe these personal operating techniques can be applied in any industry involving operations with any level of hazard. My hope is these techniques will help operators identify the signals of impending accidents, which will give operators on-going opportunities to prevent accidents and save lives. If used correctly, these personal operating techniques will enable better operational effectiveness, which will yield greater results and allow the organization or company to continue delivering valuable services or products our society needs.

Techniques for Operating Excellence 1–15

1. Develop and Maintain Risk Awareness
2. Control Risk
3. Follow Procedures (and Rules) *Thoughtfully*
4. Employ Two-person Rule
5. Identify Trigger Steps (Execution Steps with Immediate Consequences)
6. Perform Verification
7. Protect Equipment and Systems
8. Expect Failures (System and Human)
9. Develop Error Wisdom (Individual and Collective)
10. Use Error-Mitigation Techniques
11. Develop and Execute a Plan (for All Critical Phases of Operations)
12. Have a Contingency Plan
13. Preserve Options During Operations
14. Reduce Exposure to Hazards
15. Maintain Positive Control (When Moving Objects)

Chapter 4

PERSONAL TECHNIQUES FOR OPERATING EXCELLENCE
Part 1: The Technical/Process Side

The principles-based techniques described here are thought processes or mental attitudes that can be used in high-risk operations to reduce injuries, prevent accidents, save lives, and, simultaneously, improve operating performance. The results of these improvements in safety and performance by the individual operators across the organization will be to increase production and accomplish more missions with greater success. In the corporate world, this translates to raising shareholder value and contributing more to society.

These personal techniques are rarely considered explicitly in operational organizations, I suspect because they involve unseen thought processes or mental attitudes of individuals. Ways of thinking and attitudes are more difficult to describe and teach than knowledge or skills. When evaluating the effectiveness of the training and to qualify an operator to perform in a role, knowledge and skills are much easier to measure and evaluate than mental attitudes. An assessor can only infer an operator's attitudes after having conversations with the operator or observing the behaviors of the operator.

Though the proper attitudes are harder to describe, teach, evaluate, and assess than knowledge and skills, attitudes are by far the more valuable and effective for controlling risk and improving performance. Organizations populated by leaders and operators with the best mental attitudes about risk and operations don't experience major accidents. These organizations consistently demonstrate the best performance.

The first fifteen of the personal techniques are thought processes or mental attitudes individual operators use to control risk when operating with *systems*. These fifteen, described in this chapter, represent the *technical* side of the sociotechnical control used in pursuit of operating excellence.

The second fifteen of the personal techniques are thought processes the individual operators use to control risk when working with *people*, by enhancing interpersonal relationships. Those second fifteen, described in Chapter 5, represent the equally powerful *social* side of the sociotechnical control systems.

Each technique has a description and some illustrative examples. Though each of these techniques can be used singly, you and your teammates will realize greater safety and productivity benefits when you master and apply multiple techniques.

I recognize the list of techniques is long. Thirty techniques may be too many to memorize, certainly all at one time. As it is with many other human endeavors, if you find some of these techniques particularly relevant and useful in your work, you will learn them and live by them, rather than only memorize them. I hope you will identify many more examples in your line of dangerous work or hazardous operations. Over time, these techniques should form a way of thinking for you that seems effortless, continuous, and automatic. I am confident you will operate more safely and productively when these techniques become part of your normal way of thinking and operating.

1. Develop and Maintain Risk Awareness

Organizations trying to accomplish missions involving dangerous operations will be successful only if the managers manage risk and the operators control risk.

Managing risk usually involves a four-step process. The first step is to *Identify the Hazard*. When individual operators control risk, they use a similar four-step process (see *Control Risk,* the second of these thirty Techniques for Operating Excellence). But individual operators have an important prerequisite. Operators can't identify hazards reliably or control risk effectively without mastering this first technique, *Develop and Maintain Risk Awareness.*

Initially, operators should purposefully develop an awareness of risk. Over time, with the right mental attitude, operators can learn to maintain a nearly continuous high-level awareness of risk. When this technique is mastered, operators will be able to identify all hazards and reduce the consequences of any risk in every hazardous situation.

Good operators sense danger. This is a valuable ability that can save lives. At the start, the mental process of controlling risk in a new situation requires active directed attention and conscious volitional effort to identify hazards. But directed attention and conscious effort are precisely the mental processes no longer needed when performing habitual activities—even in hazardous environments. Ability becomes skill when we no longer have to think consciously about performance. Automatic processing in our brains takes over to control routine activities. When we become familiar with tasks in a high-hazard environment, not thinking about the risk becomes surprisingly easy because we no longer have to think much about the task. In dangerous work, experienced operators simply forget to sense the danger.

Most injuries and fatalities occur when people either fail to identify the hazard consciously or exhibit risk tolerance and accept the risk. The victims have become complacent to the danger precisely because they have become skilled in operating in high-risk environments.

Good managers are aware of this potential for complacency in operators. To combat complacency, managers provide operators the training and encouragement to learn this first technique of developing and maintaining continuous awareness of risk. In high-hazard endeavors, operators must never forget how dangerous their environment is.

To develop and maintain risk awareness, I use three subtechniques, anchored in the past, present, and future of operations for every system that can harm my team or me:

A.   Search for Vulnerabilities (Learn from *Past* Operations)
B.   Maintain Situational Awareness (Sense *Present* Operations)
C.   Anticipate the Changing Shape of Risk (Predict *Future* Operations)

A. Search for Vulnerabilities (Learn from *Past* Operations)

Operators and managers should understand, as best they can, the vulnerabilities of any system that is being operated. Vulnerabilities are uncovered by studying past operations. There is no airplane, automobile, boat, lawnmower, or any other system I have ever operated without trying to understand fully how that system could kill me. Vulnerabilities in the system can include any weaknesses of the hardware, software, and people operating the system. Among the best ways to understand vulnerabilities are researching design manuals, drawings, manufacturer's reports, and instructions;

reading lessons learned and historical incident reports; and talking with experienced operators, engineers, and the manufacturer.

I try to maintain an awareness and understanding of all vulnerabilities in general, before I operate any system. Knowing when people and systems are vulnerable can create success rather than tragedy. People in an operational organization are vulnerable during shift handovers or transitions. Systems in operation are more vulnerable during startup or when experiencing transient operations or mode changes. In the best organizations, operators will increase their vigilance at these times to prevent accidents.

Airplanes with only one engine are more vulnerable than multiengine airplanes. At every second during flight, good pilots who fly single-engine airplanes know exactly where they will attempt to land if the engine suddenly fails. Occupants in a conference room are more vulnerable in an emergency if there is only one exit from the room. With the knowledge of this particular vulnerability, smart occupants will consider how to escape if a fire breaks out (or escape the boring meeting long before a fire ever starts).

The next time you board a boat I suggest you determine how many life preservers are aboard and how to access one of them. If the boat has an insufficient number, your life may not be preserved.

Here are some examples to illustrate the concept of *Searching for Vulnerabilities* and how this technique can result in safe operations and better long-term performance.

Example: Space Shuttle Landing and Rollout Vulnerability
Early in the history of the Space Shuttle program, we learned the landing and rollout system had a particular vulnerability. The system of carbon-lined beryllium disc brakes in the four main wheels had limited energy absorption capability. The Space Shuttle was the only air or space vehicle of its weight class designed with only four main wheels. The braking system in each main wheel was capable of absorbing no more than forty-two million foot-pounds of energy.

During the rollout on the runway, if any brake were commanded to exceed its energy limit, the fuse plugs in the braking system were designed to release hydraulic pressure and prevent further absorption of energy. A minimum consequence was loss of braking on that wheel. A more severe consequence could have been locking of the brake because of overheating and fusing, resulting in a blown tire and, depending upon the ground speed and steering inputs, reduced steering control.

Photo 15 shows the Space Shuttle *Endeavour* immediately after main wheel touchdown on STS-113. The visible smoke is from the heated rubber of the main wheels as they spin up to 205 knots ground speed on the runway surface at the Kennedy Space Center in Florida. Specifically because of this vulnerability of low energy capability in the braking system, I decided to protect the landing system hardware and limit my use of the brakes during rollout on the runway. At the time I landed on STS-113, 2 of my landings were the longest of the 111 landings in the program. In another technique *Protect Equipment and Systems*, I will present more about this example and the control I used to improve performance and prevent an accident this vulnerability could have caused.

Photo 15: Landing of mission STS-113, showing touchdown of the main wheels, part of the vulnerable landing and rollout system of the Space Shuttle (Photo Credit: NASA)

A few other examples from everyday life illustrate the benefits of searching for vulnerabilities.

Example: Vulnerability on a Single-Lane Road
Photo 16 shows one of two roads from my neighborhood. I never relax while driving on this road when cars are approaching. You can see from the picture I have few options if the opposing driver begins to drift into my lane as a result of being distracted, drowsy, or drunk. A necessary first step in staying alive on this road is to understand this vulnerability, where I have limited options to respond if another driver veers into my lane at the last second. I will present what I do about this vulnerability in the next subtechnique, *Maintain Situational Awareness*.

Photo 16: Vulnerability on a single-lane road with opposing traffic—limited options, if the other driver veers into my lane

Example: Vulnerability with Limited Visibility At an Intersection

Operators and systems are more vulnerable at transition points during operations. Examples include starting or stopping systems, changing modes of operations, or, in this case, going through an intersection while driving. You can see in photo 17, I am vulnerable as I approach the intersection. In my low eye-height car, I occasionally encounter situations where I am not able see the cross traffic at an approaching intersection. Though the traffic light in the photo is green for me and I have the right to enter the intersection, I will not enter unless I can verify nobody is running the red light and is about to T-bone my car. Recognizing

Photo 17: Vulnerability with limited lateral visibility of cross traffic as I approach this intersection

this kind of vulnerability is important to me, and so far I have never died while driving through an intersection.

Example: Vulnerability in a Hotel Hallway
Notice photo 18 and consider why the exit sign is located near the floor. People were vulnerable in hallways that were half filled with smoke when exit signs at normal eye height became obscured. Installing the signs at a low height reduces this vulnerability in future evacuations.

When I saw this sign in a hotel in San Antonio, Texas, I was reminded of another, more dangerous, vulnerability. Even if I don't see smoke, I will still crawl near the bottom of the hallway to evacuate from a fire. Invisible, lighter-than-air, toxic fumes may be filling the top half of the hallway, waiting for an opportunity to kill me.

Photo 18: Vulnerability in hotel hallway mitigated with this thoughtful positioning of the exit sign

B. Maintain Situational Awareness (Sense Present Operations)

The previous subtechnique, Search For Vulnerabilities, concerns researching the past and understanding the history of a system and its weaknesses. *Maintaining Situational Awareness* is about staying in the present and sensing everything around us in the operation. Situational awareness is a common term used in aviation to describe how well the pilot has interpreted the current state of operations of the airplane and its environment, including any potential hazards. Through intensive training and operational practice, humans can develop a heightened ability to remain acutely aware of the hazards surrounding them, a valuable skill during critical phases of operations with high danger. Identification of hazards is best accomplished with directed attention and conscious volitional thought processes. The techniques and skills required must be continually practiced. For a more detailed explanation of the skill needed to maintain awareness, see the technique *Be Mindful During Operations.*

The power of the mind is sometimes underappreciated. Through directed effort of attention, with top-down, executive-level cognition, pilots or operators can precondition their thought patterns to respond more effectively to external stimuli. An example is the intensive training astronauts receive in simulators before flight. Through a heightened state of awareness developed with training, flight crewmembers can work together to identify minor anomalies before they become major malfunctions. The crew can sense and respond quickly and correctly to small changes in gauges that indicate preliminary problems during the powered ascent phase of the rocket's launch. Or they can take immediate corrective actions when a half-million-pound-thrust main engine fails, yet completely different corrective actions if the same engine fails two seconds later after crossing an energy boundary in performance. The best crews with the best awareness will diagnose and correct a problem before the on-board computers have recognized and annunciated the failure.

Example: Maintaining Situational Awareness in Time-critical Operations
In the previous subtechnique, photo 16 showed a particular vulnerability on a single-lane road with few options if the opposing driver begins to drift. Photo 19 shows the view of the same road through the passenger's windshield. With limited options (constrained by the concrete abutment in the ditch), and little time to react if the opposing car begins to drift into my lane, I must be acutely aware of the velocity vector of the 3,000-pound hazard approaching at high speed.

Photo 19: Maintaining situational awareness while driving with a known vulnerability and limited options—a valuable skill on a single-lane, shoulderless road

At this time in my life, nothing is more important than perceiving any subtle changes in direction, particularly with the tires generating the steering forces for the approaching car. By properly maintaining situational awareness, I will be able to sense more quickly if the car begins to drift toward me, which will increase the time available for me to react and save lives in both cars.

Unfamiliar or stressful operations can challenge the operator's ability to maintain situational awareness. Cognitive ability decreases when the brain is overloaded with new sensory information or elevated stress. During any operation when things begin to go wrong, at the very moment when an accurate understanding of the situation would be most valuable, operators can quickly lose situational awareness unless they have been trained to manage the elevated stress.

Example: Failure to Maintain Situational Awareness During Stressful Operations
A case of a failure to maintain situational awareness is documented in photo 20. The picture was taken moments after a vehicular accident on a hurricane evacuation route the day before Hurricane Rita hit Houston in September 2005. Tensions were high on this day, only one month after Hurricane Katrina devastated New Orleans. More than two million motorists made the decision to leave the Houston area, and many were stranded at least twenty-four hours on designated evacuation routes when they became gridlocked in one of the largest evacuations for a hurricane in US history.

In the situation shown in the picture, a driver stopped his silver minivan behind the tractor-trailer, which had barely entered the intersection. A police

Photo 20: Vehicular accident on hurricane evacuation route in Texas before Hurricane Rita, 2005

officer was stationed in the intersection, just to the right of the picture, and was commanding the tractor driver to exit the intersection in reverse. Either the policeman or the tractor driver, or both, failed to maintain proper awareness of the available clearance behind the tractor-trailer, which backed into the stationary minivan. The apparently irate car driver's left hand can be seen in the photo gesturing forward.

If you need to operate your vehicle in a similar situation, you should increase your level of awareness deliberately. Since an evacuation event is both stressful and unfamiliar to most drivers, you should expect an increased likelihood that other drivers will lose situational awareness.

Example: Failure to Maintain Situational Awareness with Limited or Conflicting Sensory Inputs

This example shows how important maintaining situational awareness is in high-performance naval aircraft. Photo 21 shows the results of a night rendezvous attempted by two F-14 Tomcat aviators. The task during aerial rendezvous is to decrease the relative distance, by controlling the rate of change of speed and direction, and eventually to stabilize in close formation with the lead aircraft. In this fatal accident, the trailing F-14 was attempting to join the lead F-14, shown damaged in the photo. After the airborne collision, the trailing F-14 and pilot were lost at sea.

Performing an aerial rendezvous is much more challenging in the darkness of night, with limited visual cues and almost nonexistent depth perception. Without a visible horizon, the potential for spatial disorientation or vertigo is high. The task can be completed safely but only by maintaining an accurate assessment of

Photo 21: Results of a night rendezvous after a joining F-14 Tomcat collided with the underside of the engine compartment of this F-14, which was leading the formation

the velocity vector of both airplanes relative to the surface of the earth and each other. To prevent such deadly accidents, pilots must master this subtechnique, Maintain Situational Awareness, even during a night rendezvous with limited visual cues.

One of the keys to staying alive in hazardous operations is perfecting the ability to maintain situational awareness and, if awareness is temporarily lost, to quickly regain an accurate assessment and awareness of the operational situation.

In an earlier section, I mentioned one of the great evolutionary advantages of the human brain is its ability to perform practiced tasks with learned subroutines using automatic control. This is the upside of becoming skilled. The brain can devote cognitive attention to higher-level executive thoughts. We normally don't need to think about body positioning when we walk—unless we are walking near a spinning propeller on the flight deck of an aircraft carrier. But there is a downside to becoming skilled. When operators work routinely in hazardous situations, they quickly learn to normalize the hazards. They can become skilled *and* complacent. In dangerous environments, inattention to routine tasks can kill.

Example: Visible Indications of Complacency and Not Maintaining Situational Awareness

In hazardous operations, the front-line supervisor is the most influential person in a position to combat potential complacency in an operator. As soon as possibly complacent behaviors are observed, the most effective supervisors will provide immediate, and usually strong, feedback to the operator. Improvement in behavior usually occurs just as quickly.

But to be effective, the supervisor must be attuned to any signs of complacent behavior from operators in hazardous environments. To illustrate what a supervisor might look for, this example shows the visual indication in the body language of an operator who might not be maintaining proper situational awareness of impending hazards.

Here's the background. Photo 22 shows an A-7 Corsair during an arrested landing aboard an aircraft carrier. Just before touchdown, the tailhook on the airplane snagged the arresting cable. At an initial speed of 135 knots, the one-inch-diameter steel braided cable was pulled from spools on both sides of the landing area. In the next two seconds, tremendous hydraulic pressure will slow the unwinding spools, and the cable will transfer the rotary deceleration to tension force, which will drag the 60,000-pound airplane to a full stop against the opposing thrust of the engine screaming at full power. On occasion, the tension force and vibrations will cause the

Photo 22: An arrested landing aboard the aircraft carrier USS *John F Kennedy*. On rare occasions in the Navy, similar one-inch-diameter steel braided cables have snapped.

cable to snap. When it breaks, the cable becomes a giant weed whacker able to cut people in half who fail to time their jump over it.

Cable failures are rare. When they do happen, rules, policies, and procedures won't save the flight-deck personnel. The only way to stay alive is to have already understood the vulnerability and to have maintained complete awareness of the situation on the flight deck, with its numerous hazards. If a sailor is complacent during the two seconds of any carrier arrested landing, that sailor may not live through the next two seconds.

When a front-line supervisor sees behaviors indicating a sailor might be complacent and may not be maintaining awareness of the hazards, immediate feedback is given to change work habits and improve performance. Now look at photo 23. Do you think the sailor in the dark shirt is properly maintaining

Photo 23: The body language indications of a sailor (in the dark shirt) who may not be maintaining situational awareness of the hazards of the impending arrested landing

situational awareness of the impending hazard? What would you do if you were this sailor's supervisor?

When lives are at risk, the best leaders will *change* the person or change the *person*. New behaviors or a new job are the only choices.

This subtechnique of maintaining situational awareness can be useful also in nonhazardous, more common situations we might encounter during business meetings or social situations. Through practice, we can train our minds to stay in the present and be more sensitive to our external surroundings. With increased awareness of our local environment, we can improve our self-control over our internal emotions. By being more attuned to the levels of stress in meetings or social situations, we can train ourselves to anticipate and prevent impulsive reactions and reduce the possibility of emotional hijacking in stressful situations when tempers begin to flare.

C. Anticipate the Changing Shape of Risk (*Predict* Future *Operations*)

Of these three subtechniques, the most difficult to master is *Anticipate the Changing Shape of Risk*,[16] but this one likely represents the greatest potential to prevent accidents. Because the operating conditions are continually changing, the risk is never static. If operators think they understand the current risk, they will be surprised (at best), injured, or killed (at worst), when they realize their knowledge of the risk was outdated. Operators must learn to recognize subtle differences in the operating environment, systems, at-risk situations, or any other conditions and then anticipate how the risk will change.

This valuable ability to look into the future to prevent accidents requires mastery of both sides of any operational issue, the technical/system and social/human aspects. Accurate prediction of changing risk relies on deeply understanding the technical elements of operations, including how the systems function, what vulnerabilities exist in the design, and how the equipment will respond to changing conditions.

Accurate prediction of increasing risk also requires a deep understanding of the social elements of communication, relationship-building, judgment, intuition, experience, human values, human factors, decision-making, emotional intelligence—with holistic, systemic, nonlinear thinking in a supported climate of organizational learning.

---

16   Deb Grubbe, from DuPont, used the phrase "anticipate the changing shape of risk" when she was providing expert testimony to the *Columbia* Accident Investigation Board in 2003. This single phrase changed my way of thinking about controlling risk and altered my course on a continuous, never-ending journey of understanding how to prevent accidents, while improving operating performance and organizational productivity.

In the spaceflight environment, resources are being consumed continually. This alone increases risk as the mission progresses. Additionally, flight and life-support systems are being operated, which increases wear and decreases fatigue life of individual parts. System component failures were common and expected on all six of my space missions. Risk is always increasing in hazardous endeavors, even during normal operations.

In the oil-and-gas industry, wells are drilled deep below the earth's surface into reservoirs of hydrocarbons. During the production phase, after the well has been completed, hydrocarbons are produced and transported away for refinement and eventually delivered to consumers. In all these operations—exploration, drilling, production, transportation, refinement, delivery—risk is always changing.

Even during normal production with expected operations and no failures, the conditions in the subsurface reservoir are continually changing. As hydrocarbons are lifted to the surface, the down-hole pressures, temperatures, and chemical mixtures are changing. Each of these variations will change the risk profile in the systems and equipment used to process the produced hydrocarbons. The engineers and operators must stay ahead of the changes and adjust the operational systems, equipment, and processes as necessary to control the risk. When they don't, risk increases.

An example may illustrate the unique aspects of this third subtechnique and its importance in developing risk awareness.

Example: Anticipating Changing Risk in Operations
In anticipating the cardiovascular deconditioning expected during weightless conditions on orbit and the extreme physical rigor encountered when returning to Earth's gravity, astronauts try to be in peak physical condition before liftoff.

Photo 24: A particular hazard at the gate on the fence line at the training center

To build the required fitness, running was one of our common regimens during preflight training.

Photo 24 shows me on my favorite route for daily runs on the fence line at the people gate into the training center. I didn't see a particular hazard at this gate for several years until I started thinking about this subtechnique, Anticipate the Changing Shape of Risk.

The hazard is hard to see in the previous picture. Photo 25 shows the same hazard from a different perspective. The concern is the structural integrity of the heavy, rusty spring and, in particular, how and where the stored energy will be released if the spring suddenly breaks. Consider the positioning of the spring relative to my eyes and jugular vein.

Photo 25: Hazard from a different perspective. What will happen if the rusty spring suddenly breaks? When, during its operation, is it more likely to break?

The salient point of this example is learning to anticipate *changing* risk. With experience and training, you can tune your senses to identify hazards such as the stored energy in a rusty spring. To Anticipate the Changing Shape of Risk, the relevant question is, *when* is the spring more likely to fail? It is not when the spring is sitting dormant under static conditions, when no one is using the gate. It is more likely to fail when someone's eye or neck is near the stored energy, specifically because that person is opening the gate and creating the stressing forces in tension and scraping of metal against metal. The very act of operating a system, normally or in transition, can change the risk profile and increase the danger to the operator.

This first technique, Develop and Maintain Risk Awareness, is one of the more powerful of the Techniques for Operating Excellence for controlling risk. Through mastery of its three subtechniques,

A. Search for Vulnerabilities (Learn from *Past* Operations)
B. Maintain Situational Awareness (Sense *Present* Operations), and
C. Anticipate the Changing Shape of Risk (Predict *Future* Operations)

operators will be well prepared to execute the other techniques effectively to prevent accidents and improve performance.

2. Control Risk

After developing and maintaining a deep awareness of risk, the next important Technique for Operating Excellence for each operator who works in dangerous environments is to *Control Risk*.

Controlling risk at the operator level is similar to managing risk at the management level, with some differences, as I will explain. Both are accomplished with a four-step process: (1) identify hazards, (2) assess risk, (3) implement controls, and (4) assess the effectiveness of the risk control (or risk management) process.

When implemented correctly, the risk management process results in reducing the collective operational risk to the organization below certain acceptable levels. As long as the operational risk is calculated to be lower than the acceptable level, the organization retains its privilege to operate, from both regulators and society. Risk management is an organizational process that has worked well for many years to protect the organization as a whole, even though individual operators may be injured.

Individual operators think about risk differently than managers. Operators, who face danger in their daily work, must control personal risk to prevent personal injury. The same four steps managers use in the collective organization to manage risk are the basis of controlling risk at the front line, although individual operators conduct the steps somewhat differently.

A. Identify Hazards

The first step in controlling risk is to *Identify Hazards*, which is a skill that operators should develop and continually improve to prevent accidents and enhance performance. They can acquire the skill through training and practice, and experienced operators should teach the apprentice operators how to identify hazards.

Some hazards are easy to identify, and some are not. Here are some examples.

Example: Hazardous Driving Conditions in Alaska
Spotting the driving hazards in photo 26 is easy.

Photo 26: Hazards of driving in winter conditions in Alaska are obvious.

Example: Visibility of Hazards
Some hazards are more difficult to identify. If you were on the inside of this high school building in Scotland (see photo 27) and needed to exit quickly, identifying the hazard might be challenging. The degree of difficulty depends on what indications are on the inside of the door.

Photo 27: Building in Scotland with an egress hazard, which may not be obvious from the inside

Example: Nearly Invisible Hazard in Subsiding Ground

Some hazards are nearly invisible. In this example, a hazard was not identified, and an employee was critically injured. A steam line was leaking at a refinery. Inside the pressurized line, the steam was 240°F. As the superheated steam exited the leaking line and encountered normal atmospheric pressure, the steam immediately condensed to water near its boiling point at 212°F. This scalding water began to sink into the ground in a small area under and near the leaking line, leaving little evidence of the hot water from the leak.

An employee was walking the area in preparation for the repair work, which was to be conducted later that day. As he stepped into the area near the leak, the soft ground suddenly gave way, and his legs were submerged in the newly formed pool of near-boiling water until he climbed out. Consider how difficult it would be to identify visually the hazard of ground that appeared solid yet had been made unstable by hot water.

Regardless of the challenge, the organizational leaders must help their workers learn to identify hazards, visible and invisible.

Example: How the Mind Works

Operators primarily use their sense of vision to identify hazards in the field. Steven Pinker explains how our visual system functions in his book *How the Mind Works*. In his explanation, Pinker cites a demonstration developed by psychologist Anne Treisman to illustrate automatic and directed attention, from unconscious and conscious processing in the brain's visual system.[17]

To illustrate Treisman's demonstration, I created diagrams 1, 2, and 3, as shown below.

In diagram 1, I don't even have to explain the rules of the mental game. If your visual system is capable of distinguishing the colors, your brain automatically and immediately identifies the symbol that has a different color.

In diagram 2, again, I don't need to specify the rules; your brain automatically and immediately identifies the symbol with a different shape.

Now look at diagram 3. If I ask you which symbol is the wrong color, or which of the colored symbols has the wrong shape, your brain takes much longer to identify the correct answer. Your brain will not identify the answer automatically or easily. Directed focus of attention is required, and you must control your visual system and thought processes to methodically look at individual symbols and decide whether the color or shape is either appropriate or inappropriate. Can you identify which symbol is the wrong color, or which of the colored symbols has the wrong shape?

---

17   Steven Pinker, *How the Mind Works* (New York: W. W. Norton & Company, 2009), 140–141.

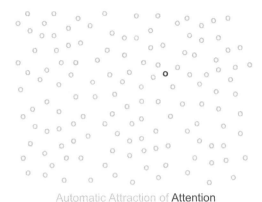

Automatic Attraction of **Attention**

Diagram 1: Demonstration psychologist Anne Treisman developed to illustrate automatic and directed attention, from unconscious and conscious processing in the brain's visual system

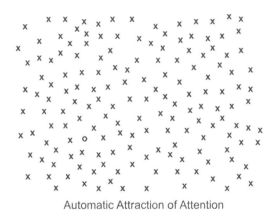

Automatic Attraction of Attention

Diagram 2: Demonstration psychologist Anne Treisman developed to illustrate automatic and directed attention, from unconscious and conscious processing in the brain's visual system

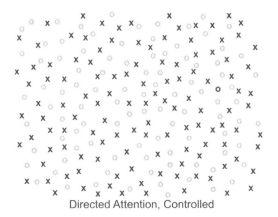

Directed Attention, Controlled

Diagram 3: Demonstration psychologist Anne Treisman developed to illustrate automatic and directed attention, from unconscious and conscious processing in the brain's visual system

With the illustrative examples from Treisman's demonstration, we can understand operators will be able to identify some hazards easily and automatically. Large leaks of red-colored hydraulic fluid pooling on a white floor will be obvious, even to inexperienced operators. However, if the evidence of a leak is small or camouflaged by its environment, identification of that hazard might require conscious volitional effort by operators with relevant experience. If you want your operators to identify hazards reliably, you must train your operators and give them the knowledge, skills, and proper attitude to use their senses effectively to perceive the subtle indications of any potential hazard.

Try a simple method my colleague John Putnam used in the oil-and-gas industry to train operators to identify hazards. In an exercise he called "Hazard Hunt," he simply projected on a large screen a picture of an area of a refinery or drilling platform. The operators began to identify any hazard they saw. Individually, some operators could see more hazards than others, and many operators were silent initially. Collectively, they learned from one another, and after several months, each operator became highly skilled in the art of recognizing hazards.

After this first step, Identify Hazards, the operators should be concerned with conducting operations in a way that will eliminate the hazards or reduce personal exposure to the hazards. Deciding how to do this brings us to the second step in the technique Control Risk.

B. Assess Risk

After identifying the hazards, the operator must *Assess Risk*. This is identical to the second step for the collective organization when managing risk, but the operator thinks about risk assessment differently than the managers. Managers assess organizational risk by calculating the numerical product of probabilities (frequency or likelihood of occurrence) and severity of consequences, including how many people might be injured.

Operators assess personal risk by using a more personalized and subjective method: Will I or anyone on my team be injured? If so, how badly? How can I prevent an accident?

The way an operator assesses risk is biased by his or her individual attitude regarding risk. Notionally, risk attitude is the personal ratio of risk perception to risk propensity. The operators with the better risk attitude have a higher perception of risk and a lower propensity to accept risk.

In my opinion, operators in various professions generally devote insufficient mental effort toward conducting a high-quality assessment of personal (or organizational) risk. My observations have led me to believe many operators think more about doing the work and getting the job done than the possibility of having accidents. Their perception of the risk is low. When these operators do consider risk or accidents at all, they generally think, *it just won't happen to me.*

Some operators like the challenge of accepting risk, and their propensity for risk is high. These operators are often overly confident, and they don't think they will suffer injuries. Accepting risk can be intoxicating for many people, especially when the risk is calculated under an illusion of overconfidence and improbable accidents.

In chapter 2, I explained my personal attitude toward risk. I always try to consider the worst consequence that can happen to my team and our equipment, the criticality or importance of the mission, and our ability to control the hazards. If I can control the hazards, I will attempt the mission. But I must anticipate the worst long-term consequence that can happen, and I expect it *will* happen if I fail to control the hazards. In this failure scenario, I consider the importance of the mission relative to the risk. The more important the mission, the more willing I am to pursue it, while accepting the higher risk, but the more tightly I attempt to control the hazards. If I, or the organization, can't control the hazards, I will not attempt the mission unless it is so important I'm prepared for it to be my final mission.

The best operators have the best risk attitude with the highest ratio of risk perception to risk propensity. Throughout their careers, operators should try to improve their risk attitude by enhancing their perception of hazards and reducing their personal propensity to accept risk, commensurate with the importance of the mission and their ability to control the hazards.

Example: Risk Assessment Scenarios
How would you assess the risk in the situation depicted in photo 28? You can see the motorcycle rider is not wearing a helmet. Not apparent in the picture, the rider

Photo 28: How would you assess the risk? (example 1)

was not wearing hearing protection, though the noise level of the motorcycle engine seemed abnormally high to me. I can only infer from the carton of cigarettes on his handlebar, the rider may be a smoker. Shortly after the picture was taken, the rider received a cell phone call. After passing through this intersection, he stopped in the center turning lane of the highway—that is, in the middle of the road—to take the call. At least, he didn't conduct the phone conversation while in motion.

What about the risk in photo 29? Incidentally, this is the other side of the same intersection depicted in photo 20 where the minor driving accident occurred. The photos were taken minutes apart, during the evacuation from Hurricane Rita. Presumably, the men in this picture were under the direction of the police officer (unseen) in the center of the intersection and were tasked with installing the temporary cover over the stop sign.

As you assess the risk for this task, think about the obvious issues, such as the small size and limited stability of the ladder, the absence of personal protective equipment, and ergonomics. The man appeared to have significant difficulty installing the cover, which was not big enough. Also, I thought about the hazards created by motor vehicles passing through the intersection near the worksite. And I wondered about the purpose of the task. Why was the cover needed? What questions would drivers need to process when they arrived at an intersection that had a partially covered stop sign, a functioning traffic light, and a policeman on-scene giving directions? Later, if the policeman departed, how would drivers respond to the partially covered sign? Would you be comfortable with any part of this task if you were responsible for assessing the risk?

Now look at photo 30. The truck was moving when I took the picture. I'm fairly confident no risk assessment was conducted. Do you assume anyone is thinking

Photo 29: How would you assess the risk? (example 2)

Photo 30: I would have assessed the risk differently
in this operation with a moving truck (example 3)

anything about risk in this situation? Who is responsible for the safety of the people
in this picture?

C. Implement Hazard Controls (Eliminate or Reduce Exposure to Hazards)

After identifying all hazards and assessing the risk, the next important step for the
operator is to put in place control mechanisms to eliminate exposure to the hazards, or
at least to reduce exposure to an acceptable level. Though this is the same next step for
the organization in managing risk, the operator has fewer options for implementing
mechanisms to control risk.

We can use the same Hierarchy of Controls listed at the end of chapter 1 to
organize the categories available to an individual operator. Although all seven
categories of controls are valuable in the organization, the operators must rely on the
managers to initiate the more effective controls, as I will explain:

- **Hierarchy of Controls** (Hazard and Error Defenses). Seven defenses that
  can be used to reduce the consequences (and number) of hazards and errors
  in the organization, in decreasing order of effectiveness and cost, are:
  A. **System Design** (Hard defenses—can change the hazard)
     1. *Design Out*: Change the design of the system to eliminate
        the hazard or possibility of error. An example is the design
        incompatibility between the nitrogen hoses and the oxygen inlets
        in the breathing systems for patients in a hospital. Because the two
        different types of connectors don't fit together, human operators
        can't make the fatal mistake of connecting a nitrogen supply to a
        breathing system.

2. *Engineer In*: Insert redundant subsystems or interlocks into a system. Redundancy or interlocks, when engineered into a system, make committing an unintentional mistake less likely.

3. *Guard Against*: Install physical barriers or preventative locks in a system. Physical fences around dangerous hardware are intended to keep people away and reduce the potential for injuries.

4. *Constraints and Affordances*: Use good Human Factors design principles in the system. As a simple example, panels on doors afford pushing, and handles afford pulling. When installed, these design elements influence human behavior resulting in fewer errors.

B. **Administrative Rules and Procedures** (Soft defenses—will not change the hazard)

5. *Warning*: Install and use warning systems. Visual or aural alarms in a system will alert the operators before an error is made or consequences occur and before the system exceeds the limits of specified performance.

6. *Training*: Train the operators to understand and follow rules and operating procedures. Remember, the rules and procedures are not the controls; risk is controlled only when the operators actually *use* the rules and procedures. Training and influencing are the mechanisms the organization uses to effect control. Rules and procedures are intended to influence personnel to operate the system within design limits and to reduce errors. Training is required to reinforce the proper use of these rules and procedures.

7. *Personal Protective Equipment*: Require the proper use of personal protective equipment. Though protective equipment will not change the hazard, the proper use of the equipment can protect people by reducing the severity of the consequences of a hazard that becomes an adverse event (an accident).

In priority order, from most effective to least effective, the first four of the seven control categories represent hard defenses and are different ways to design or redesign the system. The last three controls represent soft defenses. These are types of administrative controls that don't change the hazards but can reduce the likelihood of an adverse event—and can reduce the severity of consequences should an accident occur.

All seven of these controls are available to managers when they manage risk. But when it is time to control risk, operators don't have much ability to control the first four defenses. Though operators certainly benefit from these hard defenses, they can only change or control the third one. Operators don't have the luxury that managers have of *designing out* or *engineering in* changes to the design of the system to eliminate the hazard. They must use the systems as designed. Operators can implement controls to *guard against* the hazard by installing, monitoring, and enforcing compliance with physical barriers or locks in a system. Lastly, in the list of four hard defenses, the operator can't change the interfaces of human factors in the design of the system. The operator can only understand human fallibility and the potential for making errors.

This leads us to the category of controls most available to the operator, the administrative rules and procedures, which are the soft defenses. In the hierarchy of controls for the collective organization, these three soft defenses are the least effective. But for the individual operator, who is strapped into a seat on the front lines of danger, these soft defenses can mean life or death.

All three of these soft defenses are important to the operator. With little ability to insert or change the hard defenses, the soft ones become the last line of defense. Of course, the operators must understand and use the warning systems. There is no excuse for failing to use the proper protective equipment. By far, the most important of the soft control mechanisms for operators is to learn from the training and operate the system correctly. Operators who understand the technical system and operate the system properly can save lives. But operating excellence comes from so much more than only training.

In hazardous endeavors, every operator should be on a life-long journey to strive for operating excellence. Each operator should possess a deep personal motivation to learn more than the training provides. From my first day in ground school training to become a naval aviator until the present, I have always believed in the proper knowledge, skills, and attitudes—the KSAs—to keep me alive. If I had the technical *knowledge* of the airplane systems and flight operations, the physical *skill* to control the systems, and the disciplined mental *attitude* to strive for operating excellence, I could accomplish more missions, more successfully, and live to do it again another day.

Example: Fall Protection Not Used

Here is an example of workers who have not implemented the proper controls. They may have identified the obvious hazard of falling from a height, and they may have assessed the risk purposefully and precisely. But then they didn't implement the controls. Three of the five workers on the upper level in photo 31 do not appear to be wearing harnesses; none of the five appear to be tied off properly. In a valid risk assessment, the level of a worker's skill should not be considered as a control

Photo 31: Controls available (fall protection) but not used by each worker

mechanism to mitigate the hazard of falling from heights. Even the most skilled workers may fall eventually. Is today that day?

D. Assess Effectiveness of the Risk Control Process

The personal technique Control Risk has one final and important step. This is similar to the final step in the organizational technique Manage Risk. The individual worker and the supervisor should routinely assess the effectiveness of the first three steps in the personal process of controlling risk.

A routine assessment of effectiveness is valuable for many reasons. In hazardous environments, risk is dynamic. New hazards may have emerged since the original hazards were identified. The hazards may be the same, but the conditions of operations may have increased the risk. The controls in place may degrade over time. Even if the decisions were initially valid regarding hazards, risk, and controls, the conditions will change over time and new decisions must be made. Additionally, there are benefits to assessing the effectiveness of controlling risk specifically due to the fallibility of humans. Maybe not all of the hazards were identified properly; maybe the risk wasn't assessed well; maybe the controls were not in place for every worker, or success was breeding complacency and some workers stopped using the controls.

Based on my observations, most injuries occur in hazardous environments after individuals fail to *Identify Hazards* (step 1) or fail to *Implement Hazard Controls* (step 3) because of being risk tolerant or complacent. By conducting step 4 properly, managers and operators can identify these failures before an accident.

One final thought about this fourth step, *Assessing Effectiveness of the Risk Control Process*. This is the important step for learning. Whether the risk was controlled well or not, a routine assessment will help individuals learn. An after-action-review conducted today can save a life tomorrow.

Example: Multiple Hazards Exist, Ineffective Controls In Place
This example illustrates a situation with multiple hazards. In the left picture of photo 32, you can see how the ladder is configured. Would you want to ascend or descend using this ladder? Would you feel comfortable getting on or off at the top? Do you think sufficient barriers are in place on the ground to protect people from falling objects?

In the right photograph, taken from inside the building, you can see the same ladder. A worker is near the bottom after descending from the top. If you look closely, you can see the worker is carrying his fall-protection harness in his right hand. Fortunately, he did not fall. Also fortunately, he did not drop his harness on someone below. But unfortunately, his lucky success that day may reinforce his behavior to control risk just as poorly in the future.

Photo 32: Multiple hazards exist; controls are available but are not being used.

Example: Failure to Assess the Effectiveness of Risk Control
A supervisor asked me to offer my opinion on possible control barriers to protect his workers from a particular hazard. In his workshop was a large, center-pivot milling machine capable of rotating at high rpm a three-foot-diameter, heavy block of metal with a hollow central bore. A central auger milled the unsculpted block from the inside into a large component with cylindrical symmetry for use in an industrial facility.

The operator's task was to properly position the large metal block on the rotating turntable, start the machine turning, verify proper positioning of the block and the auger by looking down into the central axis of the spinning block, and initiate the automatic operation of the auger. When the milling process had been initiated, the operator would back away from the spinning block but continue to watch the operation for anomalies that might require a quick shutdown.

When positioned next to the spinning turntable of the machine, the operator was standing on a platform raised four or five inches above the shop floor. The supervisor was concerned one of his operators might stumble off the platform and twist an ankle while backing away from the rotating machine. The supervisor was soliciting my opinion regarding a possible fence or barrier to prevent an injury to an operator who was walking backward toward the edge of the raised platform, while being especially attentive to the operation of the machine.

I asked the supervisor for a photograph showing the machine and the edge of the raised platform. From the photograph (not shown), I could see why the supervisor was concerned. To his credit, the supervisor was showing great care for the safety of his workers by trying to improve the operating conditions. He was engaged in a process (Implement Hazard Controls, step 3) of trying to identify a barrier to install that would control the risk of the particular falling hazard.

However, I could see in the photograph two other, greater hazards, which the supervisor did not identify or address with me. These omissions indicated his assessment methodology (Assess Effectiveness of the Risk Control Process, step 4) wasn't being conducted properly. In the photograph, the operator was shown standing on the platform near the spinning block with his shirttail not tucked into his pants. This oversight was a violation of a safety rule for the operation of this rotating machine. As the operator was leaning over the block, he was exposed to the hazard of his untucked shirttail becoming tangled in the machinery, which could have pulled the operator to his death. Such tragic events have happened before on similar center-pivot milling machines. Additionally, in the photograph the supervisor captured and sent to me, the operator was shown without hearing protection, while standing directly under a sign indicating hearing protection was required.

Hazards can be overlooked. Humans are fallible. This fourth step, Assess Effectiveness of the Risk Control Process, provides the best opportunity for the operators and their supervisors to learn how to control risk effectively. With a proper and routine process of assessing effectiveness, operators and supervisors can learn from their mistakes and successes, and improve operations.

3. Follow Procedures (and Rules) *Thoughtfully*

All humans make mistakes. One of the best ways to reduce errors is to use procedures when conducting complicated operations. In the middle of the twentieth century, the accident rate in the field of aviation was much higher than it is today. Analyses showed pilots were making errors causing a high percentage of the accidents. What saved the industry—and many lives—was the development and use of procedures and checklists. Human errors were reduced, in quantity and severity of consequence, and the number of accidents decreased.

The purpose of procedures and checklists is to help the operator remember all the steps required to conduct a specific task or activity. Operators need procedures when conducting complicated operations. Usually, procedures contain a complete list of the steps required, which may be written in full sentences or paragraphs. The procedures may also include ancillary information between steps, such as explanatory rationale, advisory notes, or safety cautions and warnings.

When the operating tasks are simpler, checklists can be used more efficiently. They contain the complete list of steps required to conduct a specific task or activity, but the steps are generally written in abbreviated form, without amplifying information. The items in a checklist may simply be the names of each switch or pushbutton in the cockpit with the desired switch position, such as on or off.

Checklists are usually reserved for routine tasks an operator conducts frequently, so the abbreviated information is sufficient. Procedures are generally used in operations conducted infrequently and, therefore, contain more information than the checklists. As an example, pilots use checklists to start an engine routinely on an airplane. Astronauts and ground controllers use procedures to conduct infrequent (and much more complicated) rendezvous operations in space.

Regardless of the content detail and the familiarity operators have with tasks, operators need knowledge and skill acquired through training and practice to use both procedures and checklists successfully.

In the aviation industry, the use of procedures and checklists is mandated by company policy and required by government regulations. More powerful than policy and regulations, though, pilots understand the value of using procedures and checklists. In modern times, conducting ground and flight operations without using procedures and checklists has become culturally unacceptable for pilots in the cockpit.

Some operators in other hazardous industries still believe that using checklists is a sign of inexperience or weakness. This erroneous belief exists even today, after the aviation industry has proven the effectiveness of checklists in reducing human errors and accidents. Pilots know experience doesn't always help. Checklists do.

By the day I left Earth for the sixth time, I had spent more hours in the simulator than every other astronaut, with the exception of John Young (who flew twice to the moon and had a twenty-year head start relative to me in flying the sim). I believe I made more mistakes in the simulator than every other astronaut, including John Young. Errors are part of being human. Making mistakes is one of the best ways to learn. For operators who work in dangerous businesses, the goal is to make fewer and smaller errors over time. Good pilots have methods and techniques to capture errors immediately and eliminate the consequences or reduce them to an acceptable level.

For the procedures to be beneficial to the organization and its operators, two conditions must be met. First, the procedures must be accurate. Second, the operators conducting the work must actually use the procedures.

Accuracy implies the procedures reflect the organization's collective wisdom of the best way to accomplish a task. Accurate procedures can be created and maintained with an effective *Procedure Development and Review Process*. But even with a good process to create accurate procedures, sometimes the environmental conditions or the systems change during the operations. Sometimes the humans conducting the Procedure Development and Review Process make errors that pass through the process without being corrected. Astronauts had a saying that summarized the issue of occasional inaccuracy in the procedures: "There are only two ways to get into trouble with procedures: not following them, or following them blindly." This is why the technique is to Follow Procedures *Thoughtfully*. Specifically how to be thoughtful when following procedures was the goal of many training sessions in the simulators at NASA.

Before the workers in any industry will choose to use the procedures, they must believe the procedures will help them succeed in their work. Unfortunately, some workers don't believe the procedures are helpful. Some workers in the oil-and-gas industry were insulted when they were asked to use procedures. They erroneously believed the managers emphasized using the procedures because the managers thought the workers were not skilled enough and couldn't do the job without procedures.

The front-line operating leaders (called supervisors or team leaders in some operations) should train and inspire the workforce to use the procedures correctly and thoughtfully.

Later in my career, I met with executives and senior managers who were frustrated because their operators wouldn't use the procedures or checklists. Usually, these managers attempted to enforce the use of procedures with "zero tolerance," by imposing a threat of negative consequences, or punishment, which would be applied to all operators who were caught operating without procedures. These frustrated managers often asked for my opinion, based on my experience in the aviation, aerospace, and oil-and-gas industries. My advice was always the same.

Two important issues must be addressed before expecting operators will use the procedures. The managers should ask each front-line leader in the organization these questions:

1. Do you think all your operating procedures are accurate?
    ("Accurate" includes being effective and representative of the organization's collective wisdom on the best way to accomplish a task or activity.)
2. Does each of your operators think all your operating procedures are accurate—and will help him or her be successful?

If the answer to the first question is no, the organization must improve the procedures before senior managers demand the operators use the procedures with zero tolerance for failing to comply. If the answer is yes, indicating the front-line leader thinks *every* procedure is accurate, the manager should still ask how that leader develops confidence in the quality of the procedures. The only acceptable way for managers (and all personnel) to develop confidence is to use an effective organizational process for developing and periodically reviewing every procedure.

The answer to the first question leads to the second. If the operators don't think procedures will help, the managers and front-line leaders should not expect procedures will be used, even under the threat of punishment.

The front-line leaders are responsible for helping the operators understand the value of using procedures. The managers are responsible for giving the front-line leaders the knowledge, skills, and opportunities to help the operators understand the value of using procedures.

Even if the front-line leaders claim to believe every operator thinks the procedures are helpful, the managers should determine how the front-line leaders develop confidence in what each operator thinks.

These questions help the managers and front-line leaders begin to understand the depth of their problem and where to focus their attention to improve the situation. The threat of punishment will not create compliance if the managers have not created a process to provide accurate procedures *and* have not convinced the operators the procedures will help them succeed. After both questions can be answered affirmatively, all operators will use the procedures willingly.

As I illustrate the use of this technique with examples, I begin with the temptation that operators (and managers) feel to violate safety rules. Among all the organization's policies, procedures, and rules, safety rules are the established standards people are most tempted to violate. As you think about why this might be, I offer two postulates to consider:

- Everyone has been tempted to violate safety rules.
- Almost everyone has violated one or more safety rules.

Example: Temptation for Violation

Here is a story I use to illustrate why personnel are tempted to violate safety rules. At NASA, pilots were limited to a twelve-hour crew duty day when flying solo in the T-38 aircraft. The rule was intended to increase safety by decreasing fatigue. As I was training for my fifth space mission, my first training event on one day was an interesting and nonstressful meeting that commenced at 11:00 a.m. Later in the afternoon, I flew from Ellington Field, in Houston, to El Paso, for a training sortie in the Shuttle training aircraft at the White Sands lakebed in New Mexico. This part of preflight training was always fun and stress-free.

After ten approaches to complete the training, I jumped into my T-38 for an uneventful and relaxing flight on a clear and calm night back to Houston. Normally, flying eastward from El Paso to Houston in one flight is easy with full fuel tanks. On this night, with the hot temperature and high-density altitude at the El Paso Airport, I could legally take off only after removing some fuel. Still, with such good landing weather, I likely could have flown all the way to Ellington Field with the reduced amount of fuel.

However, my Navy experience taught me this wasn't a good idea. Even on this kind of perfect night when emergencies are least expected, something out of my control could occur to delay my landing until my fuel reserve was too low. So I decided to drop quickly into Kelly Air Force Base in San Antonio to refuel before continuing to Houston.

Here is where the temptation to violate the rule occurs. With this slight delay for refueling, I could see I would land at Ellington by 11:15 p.m., a mere fifteen minutes past my twelve-hour limit for the length of a crew duty day. Should I violate the rule and continue on to Houston, or stay in San Antonio for the night?

On the "violate" side of the ledger for decision making, I was well rested, highly trained, and quite experienced; I had a great day, I was not fatigued, the weather was gorgeous, and the airplane's condition was perfect; and I would be only fifteen minutes over the solo limit. Incidentally, if I had an extra crewmember—even if he or she were sleeping in the backseat—I would legally be allowed to fly two extra hours! Lastly, there was almost no chance of getting caught if I continued to Ellington after refueling. A long list of arguments supported me violating the rule.

Additionally, in favor of violating the rule were the benefits of avoiding the hassle of hitching a ride to town, purchasing a hotel room, calling back to the training department at NASA late at night to cancel my next day's training events, flying to Houston in the morning, and rescheduling the missed events by squeezing them in sometime later into our already packed training schedule. Again, the devil was on my shoulder tempting me to violate the rule and continue to Houston that night.

On the "do not violate" side of the ledger, there was only the rule standing in my way, in all its rigid conservatism. Oh, and my integrity.

The point of the story and the answer to the question of why people are so tempted to violate safety rules is, almost always we can do our jobs easier, faster, and more economically if we eliminate the obstacle slowing us down, the conservative safety rule. Rarely does violating a safety rule result in an accident or injury. If we escape the bad consequence of an injury, and we don't get caught, the behavior of violation is reinforced the next time. Rather tempting.

Now flip the logic around and look at the chronology of accidents and violating rules backward in time. If an accident *has* happened and someone has become bloodied, bruised, or broken, almost always the investigators will retrospectively determine a safety rule was violated, which contributed to the accident. You might think this logic would be sufficient to dissuade anyone from violating a rule, but I've seen too much in my career that disconfirms the thought. Too many operators in dangerous businesses don't think they'll get hurt or caught, or more plainly, don't think about rules or risk.

Here's what I think about rules. Even with the best training on or off the planet, as experienced as I am, and as safe as I have been in the past, I know I can get hurt today or tomorrow, in an instant. And I really don't like pain. The rules help me avoid accidents. Finally, though, there is one consideration that rises above all others. More than a fear of being hurt, I fear injuring my integrity. So I don't violate rules.

I stayed in San Antonio that night in the Menger Hotel in the last room available, the historic King Suite, with a giant, comfortable bed (as if I were being rewarded for making the right decision).

Example: Simple Safety Rules

Violating rules often helps violators achieve goals more quickly and easily. As evidenced in photo 33, safety rules are conservative, and violators do not often suffer the bad consequences of causing damage or being injured or caught (except, in this case, by my camera). Someday they will be injured. Every day their integrity is damaged.

Photo 33: Of rules, policies, and procedures, simple
safety rules are the most tempting to violate.

Example: Critical Safety Rules

Because of the importance of safety, some rules are highly detailed and precise, often
with valid reasons. I can infer from photo 34 the intent of the rule makers may have
been to draw attention to their requirement to drive slowly by using an unexpected
speed limit. Quite possibly, they also wanted to send the implied message, "We really
mean it this time—not even a little bit faster!"

Photo 34: Specific and precise safety rule in this speed limit

Example: Confusing or Contradictory Rules

Some rules are confusing or contradictory. Operators are sometimes faced with
dilemmas. What do I do if I don't understand the rule? Am I free to choose either

Photo 35: Exit (or is it an entrance?) in a sporting goods store

meaning if a rule is ambiguous? Which rule do I choose if two apply to my situation but contradict each other?

Photo 35 shows a contradictory and confusing rule, at least, for anyone who might actually read such signs and attempt to interpret them. I suppose the sign installer in this situation was trying to indicate, "This is an entrance, so don't use it as an exit and run into our eager, cash-carrying customers." By the way, some store employees saw me taking the picture. One week later, I returned and noticed the signs were changed.

Example: Using Procedures during Rendezvous Operations in Space
Photo 36 shows me on mission STS-113 using procedures to coordinate the rendezvous operations between the flight crews on the Space Shuttle *Endeavour* and the International Space Station in orbit, and the ground control teams on Earth.

Photo 36: Using the rendezvous procedures on mission STS-113 (Photo Credit: NASA)

Conducting a rendezvous between two orbiting spacecraft is a complicated and dangerous operation. In the critical final four hours of the operation, the flight crewmembers and the ground controllers must complete hundreds of actions accurately, in a particular integrated sequence.

The rendezvous operations culminated in bringing the Space Shuttle and the International Space Station together. Each vehicle had a mass greater than one hundred tons. Though we were traveling at a speed of 17,500 mph, the relative speed between spacecraft was only one-tenth of a foot per second, with an acceptable tolerance of plus or minus three hundredths of a foot per second. The allowable lateral tolerance at the point of docking between the spacecraft, which were more than 120 feet long, was only three inches.

American engineers designed the flight-control systems well, so the task was safely achievable—but only if the operators and controllers performed their jobs without error. The best way to accomplish error-free performance during a rendezvous in space is to *use procedures*. The motivation is compelling. Not using the procedures *thoughtfully* could result in loss of the mission, or worse, loss of both vehicles and crews.

On the spectrum of possible operations, the preceding example resides on the extreme side of dangerous and complicated processes, which require operating excellence to succeed and survive. What about less dangerous or complicated operations? What about minor rules or simple procedures? What about rules that seem to be overly conservative and not necessary, especially if we are trying to accomplish important missions and don't want to be slowed down by safety rules considered trivial by some operators?

Here is a summary of my thoughts on rules and procedures. I include a short story to illustrate how I once responded when challenged by a peer who thought I was too rigorous in memorizing and following all rules.

Every rule was made for a reason. To violate any rule is to challenge its reason for being, and to dare that reason to kill you. If you don't know the reason for the rule, knowing what actions are appropriate to prevent an accident will be much more difficult. You won't see the accident coming until just before you die. If you *do* know the reason for the rule and you violate it anyway, you are indicating to everyone else who adheres to the rule that you think you are a better and smarter operator than they are. Even if you're not caught or killed, good operators will not want to work with you. I feel having good operators want to work with me is more important than accomplishing missions at the expense of ignoring or violating rules. So I follow the rules, even in situations that don't involve high risk.

Example: Peer Pressure to Ignore Minor Rules

Early in my NASA career, a rookie classmate confronted me for being overly concerned with knowing all the rules, even the "minor" ones few other astronauts seemed to care about. My peer had one of the best reputations among all astronauts, including the veterans, and he challenged me publicly for worrying too much about every rule. He and I were both new and only recently immersed in the operating culture of the Astronaut Office. I didn't yet have a sense of the cultural norms or the collectively expected behaviors for following rules, especially as viewed by the veterans. But their collective expectations concerned me less than my internal values.

I had learned an important lesson in the US Navy (and earlier from my parents) about the value of rules. As a young aviator during a squadron training exercise, I had intentionally violated an obscure and minor rule, which disappointed a respected senior officer. My intent was to complete the mission more expeditiously to "help" my squadron score higher on the exercise graded by the senior Air Wing assessors. Though our squadron did score higher, my commanding officer was disappointed because I lessened or cheapened our grade with my violation. The small rules matter as much as the big ones if we want to be proud of our operations. Additionally, I would much rather be respected by the people I respect, who see value in following rules.

Back to my original story . . . My challenger and I were both naval aviators, and we knew I could expect damage to my reputation if I backed down from his challenge. When he confronted me about always following rules that day in the Astronaut Office with others watching, he forced my hand. I wondered if some of the other astronauts might have agreed that I was obsessive about rules. So I tried to explain my rationale for knowing the rules, while simultaneously adding a challenge intended to end this confrontation in one sentence, by saying, "I'll play the game on *their* field, using *their* football, and *their* rules, and I'll still beat *you* every time." A victory is only worthwhile if the game is played by all the rules. A mission is only a success if it is conducted correctly and honorably. Once I voiced my opinion, I didn't feel the need to convince any other astronaut, rookie or veteran, about my philosophy on rules.

So far, I have explained the operating philosophy of following procedures. But what does it mean to *thoughtfully* follow procedures and rules? The meta-level process listed here describes specifically *how* an operator should be thoughtful when using this technique:

A.   Use the Procedure—Every Step

(1) Before taking action, consider each step thoughtfully:

- Understand the purpose of each step;
- Decide if the step is appropriate (that is, will the action accomplish the objective, while reducing overall risk?);
- Know how to complete the action;
- Predict the intended consequence of completing the action;
- Predict unintended consequences and know what to do if any occur;
- Begin to look at the appropriate indicators to verify the intended consequence immediately before taking the action.

(2) Take the action, as specified by the procedural step, if appropriate.

- If not appropriate, follow step B, below.

(3) After completing the action:

- Verify the intended consequence occurred;
- If unintended or unexpected consequences occur, take corrective action.

B.   When Any Step Shouldn't Be Used as Specified, Follow an Established Process

If you predict the step will increase risk, doesn't apply, can't be followed, or isn't understood, follow the established back-out process, including:

(1) Establish a safe configuration of the system (the equipment, software, and people) and verify the operations process is in a safe condition;

(2) Tell the proper authority, usually the team leader or supervisor;

(3) Support the procedure revision process.

Example: Opening a Hatch on the International Space Station

Two spacecraft dock by connecting mechanically and forming an airtight seal around two hatches, one on each vehicle. After the mechanical connection has been verified, the vestibule seal must be tested for pressure integrity. If either crew has made any errors, the penalty after opening either hatch could be catastrophic. Life is not sustainable in the unforgiving, cold vacuum of space.

Photo 37 shows an excerpt from the procedure to open a docking-system hatch on the International Space Station. Every step must be followed correctly. We can't afford to make a mistake in space. We live and die by the procedure.

Consider the previous sentence one more time. How is it possible we can die if we are following the procedure correctly? Remember, procedures are written by humans, who are generally smart, capable, experienced, and dedicated professionals. But human, nonetheless. Humans make mistakes. Whose fault is it if a critical procedure contains an error? The answer is irrelevant to an operator in a dangerous

2.106 HATCH OPEN AND DUCT INSTALL (BYPASS CONFIG)
(JNT OPS/LF1 - ALL/FIN/SPN)   Page 5 of 10 pages

	7.3	Install Docking Target Base Plate Cover. Install APAS Hatch Cover. Secure Hatch in open position with PMA APAS Hatch Standoff.
	7.4	Stow 10" Adjustable Wrench in NOD1D4_G2. Stow 1-1/2" Open End Wrench in PMA.

        8. OPENING ODS HATCH

CRT | SPEC 66 ENVIRONMENT |

        8.1 When Cabin dP/dT < ± 0.01, proceed.

CRT | SPEC 177 EXTERNAL AIRLOCK |

        8.2 $\sqrt{A/L}$ – VEST $\Delta P \leq 0.5$ psid

            Open ODS Upper Hatch per decal.

            Equal vlv (one) → OFF, cap installed

        9. REMOVING DOCKING EQUIPMENT

> **WARNING**
>
> Surfaces may be below freezing for a short time after initial ODS Hatch opening. Avoid direct contact with vestibule surfaces until SHUTTLE VESTIBULE TEMP 1,2 (two) indicate > 40° F (SM 177 EXTERNAL AIRLOCK).

ODS   9.1 For each Docking Light
  Vestibule | Disconnect cables.
                  | Install caps on outlet.
                  | Remove the locking pin.
                  | Remove Docking Light.
                  | Reinstall locking pin.

        9.2 Remove crosshairs.
             Stow lights and crosshairs as required.

Photo 37: Excerpt from the procedures to open the docking system hatch to the International Space Station from the Space Shuttle after docking

environment. It doesn't matter whose fault it is for creating the error in procedures. It is the operator, facing hazards on the front lines of danger, who is responsible for accomplishing the operation successfully.

To accomplish an operation without error when using procedures, every step of the procedure must be considered—thoughtfully—before executing the step. Using procedures thoughtfully, by employing the meta-level process I listed above, has the double benefit of preventing operator errors and identifying any errors in the procedures. On many occasions, because I took the time to think about each step before executing—with intent to understand the purpose of each step, how to execute each step, and the predicted consequence of execution—I caught myself just before I would have made an error. On a few occasions, when using this thoughtful process, I was able to identify errors in the procedure before executing an erroneous step.

Example: Opening an Internal Hatch on the International Space Station
When opening an external hatch to space, the criticality of a flawless operation is obvious. Not so obvious, but just as critical, is opening an *internal* hatch on a space station. Just as it is with watertight hatches on marine vessels at sea, it is vitally important to leave an air-tight hatch in space in the proper configuration, open and ready to be closed to provide isolation from hazards, or closed and ready to be opened quickly to provide a safe escape route. Following procedures thoughtfully is the best way to operate successfully.

Photo 38 shows the procedural steps for opening a normal hatch on the International Space Station. In space, you don't simply open a door; you follow the procedure, *thoughtfully*.

Photo 38: Procedures for opening a generic hatch on the International Space Station

Example: Procedures for Waste Collection System Operation
Here is my third example to illustrate the importance of following procedures thoughtfully. I don't know how obvious this example is to illustrate the criticality of following procedures, but I really did not want to make a mistake when operating the waste collection system, the toilet, on the Space Shuttle (illustrated in photo 39). The seriousness of following this procedure was not obvious to me until just before I needed to use the system on the first day of my first flight in space.

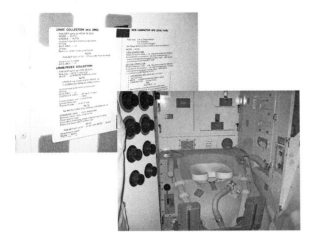

Photo 39: Procedures for operating the waste collection system (toilet) on the Space Shuttle, with a picture of the system an astronaut would like to use without making procedural errors

Before flight, I devoted insufficient time studying the procedures for the simple toilet system. I spent most of my mental effort learning everything I could about every type of dangerous failure or potentially catastrophic hazard in space, such as losing a main engine in powered flight. Until I arrived on orbit, I hadn't realized I reserved insufficient memory storage capacity for remembering how to operate the toilet. When the pressure was greatest, knowing my butt would soon be connected indirectly to the vacuum of space through the toilet venting system, I followed the procedure thoughtfully and didn't make any mistakes. But, that process took longer to execute then was comfortable. I never made that higher-level mistake again, of failing to train properly to use every procedure, even the ordinary ones.

Example: Humidity Separator Leak, STS-32

In this section, so far I have been writing about the importance of following every step in every procedure thoughtfully. I use this next example to illustrate the potential consequence of failing to use a procedure. This is the story of the one time in my space career I did not use a procedure to help me conduct a task, and I forgot to execute an important step in an emergency operation.

Several days into my first flight, my commander, Captain Dan Brandenstein, US Navy, was conducting a routine inspection of the operating condition of our Space Shuttle, *Columbia.* He looked below deck, into the volume under the middeck floor, and noticed water leaking from our humidity separator system. Photo 40 shows the result of the leak and how water adheres to objects in microgravity and forms curved shapes under the force of its own surface tension.

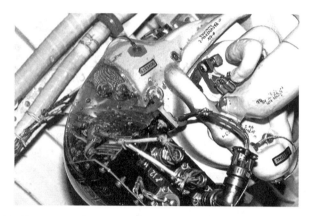

Photo 40: Free water from leaking humidity separator on mission STS-32

Dan immediately directed me to assemble the water cleanup kit. This system is essentially a wet vacuum cleaner, capable of sucking fluid into water tank D, which normally collects wastewater from the toilet. I decided to assemble the kit expeditiously and chose not to take the time to find and use the procedure. My concern was to collect the leaking water from the humidity separator before it had an opportunity to migrate onto any other electrical component that might cause an electrical short in *Columbia*. In the next few seconds, I realized the procedure would have helped me, but I made the rookie mistake of not giving the procedure a chance to help.

In my haste to connect the hose, I forgot wastewater tank D still had some residual pressure. I failed to execute the important step of relieving the backpressure using the switches on the panel shown in photo 41. As soon as I connected one end

Photo 41: Panel controlling valves in the wastewater tank system on the Space Shuttle

of the hose to the panel, a two-foot-long slug of the most foul-smelling fluid was ejected from the nozzle and began to float weightlessly across the cabin, bore-sighting directly toward the face of my revered commander. Fortunately, he saw the fetid stream coming and slowly ducked his head away from its trajectory, just in time, and the stream splattered against the side bulkhead. Unfortunately, the look he gave me was more foul than the smell of the water. He didn't say anything, but he didn't have to. Never again, on that flight or on my next five, did I fail to use the procedures.

A final thought about policies, procedures, and rules. Previously I asserted, of all the established standards, people are most tempted to violate safety rules. This is because safety rules are often conservative, and violations don't usually result in accidents. But are safety rules always conservative?

Early in my career, I heard many naval aviators complaining about safety rules they thought were overly conservative. I knew safety rules were intentionally designed to be conservative to protect all operators, regardless of their individual skill level. But, I also believed some safety rules, designed by the organization, were not conservative enough to help me stay alive and complete missions in the dynamic and dangerous environment of aircraft carrier flight operations. The following example is used to illustrate this point and to suggest operators need to do more than merely follow the organization's procedures in their pursuit of operating excellence.

Example: "Controlled Flight into Terrain" in the A-7 Corsair

The tactical mission of the single-seat A-7 Corsair was light attack, in the air-to-ground wartime strategy. A typical training flight consisted of launching from an aircraft carrier at sea, going feet-dry over land at a selected point and time, flying over

Photo 42: Land ingress on low-level navigation route to a simulated target for training in the ground attack mission

Photo 43: Low-level navigation route

Photo 44: Low-level navigation route

Photo 45: Low-level navigation route

a sequence of ground checkpoints, navigating along a route at 450 knots airspeed—two hundred feet above the ground to avoid simulated enemy radar—delivering ordnance to a ground target at a specific time, and returning to make an arrested landing aboard the aircraft carrier. Photos 42–45 show pictures I took in an A-7 after launching from the USS *John F Kennedy* in the Mediterranean Sea and flying over friendly territory to simulate the ground attack mission.

In peacetime, during the second half of the 1970s across the fleet, we were losing about one A-7 aircraft per year on these high-speed, low-altitude training flights. The incidents involved pilots unintentionally flying into the ground just after passing over one of the initial checkpoints along the route while turning toward the next checkpoint. A safety analysis was conducted, and the investigators postulated the following cause for the fatal accidents.

In the initial configuration of the A-7E model aircraft, the navigation computer had sufficient storage capacity for the latitude and longitude of only ten checkpoints. After flying over the first checkpoint, the memory space for those coordinates was no longer needed, so the pilot manually entered the coordinates for the eleventh checkpoint into the now-available memory space. The computer input panel was located outboard of the pilot's right thigh (next to the yellow rectangular lever shown in photo 46).

The investigators surmised the pilots were momentarily distracted by looking down into the cockpit to enter the coordinates, while banking toward the next checkpoint. They experienced unrecognized disorientation, loss of awareness of their velocity vector, and flew into the ground. This kind of tragedy occurred often enough in the aviation industry that accident investigators began to use the not-so-tragic-sounding label Controlled Flight Into Terrain.

Photo 46: A-7 Corsair cockpit layout. Panel for navigational computer is behind the yellow seat handle.

A new safety rule was promulgated to all A-7 pilots in the Navy. The rule essentially stated pilots were prohibited from making keyboard entries into the computer while turning at low altitude. We had to wait until we were wings level in steady flight, when maintaining awareness of the velocity vector was much easier, before making any keyboard entries.

I wasn't satisfied with the new rule. The conversations I had with pilots at the time indicated most pilots understood and accepted this new rule. I thought the rule missed the mark and provided insufficient protection to me at 450 knots, two hundred feet above the ground, a few heartbeats away from no more heartbeats. Distraction duration was much more important as a contributing factor than simply being in a turn. Surely, the turn exacerbated the tendency to become disoriented, but I could have flown into the ground just as quickly with my wings level if I were distracted for more than a few seconds. Photo 47 should solidify this point. This is a picture I took on a similar low-level navigation route, but in this case, the route was through mountainous terrain. Mountains don't care if we are turning or wings level. To kill us, the mountain only needs us to be distracted for a short period of time.

Photo 47: Low-level navigation route through mountainous terrain

My solution was to self-impose an additional limitation when using the computer keyboard. I never looked down longer than the time it took to type three keystrokes before pausing to look up and verify my velocity vector. Every time I entered new coordinates, I needed more time, mental effort, and discipline than if I hadn't self-limited the keystrokes I typed without looking up. Sometimes over level terrain, I thought I could safely type four or five keystrokes, rather than three. But, bending the rule occasionally, or allowing the rule to creep toward being less conservative, is the same as violating the purpose of the rule, which was only trying to keep me alive.

Most safety rules are conservative, but some safety rules are not conservative enough. As operators, we need to be *thoughtful* about the procedures and make sure they will protect us properly. When they don't, we need to protect ourselves by self-imposing additional limitations.

4. Employ Two-person Rule

Two people working together have a greater chance of trapping potential errors and completing a procedure or task correctly than a single person working alone. Spaceflight crews honor a rule specifying all procedures will be conducted by at least two crewmembers, whenever possible. On occasions when all other crewmembers were scheduled with independent operations and no other crewmembers were available to help, the flight controllers on the ground served as a superb substitute for a second person.

Example: Following the Two-person Rule
Photo 48 shows my pilot, Colonel James "Vegas" Kelly, US Air Force, with me on board the Space Shuttle *Discovery*. Vegas worked the two-person rule with me better than anyone in my twenty-eight-year career as a pilot. He always seemed to know exactly what I was thinking during operations. When we worked together, he was able to predict and prevent errors I was about to make.

Photo 48: Using the two-person rule with Pilot Jim "Vegas" Kelly on mission STS-102 (Photo Credit: NASA)

Example: In-flight Latch Test with Assistance from Our Ground Team
On my fifth mission, one of my scheduled events was an in-flight test and demonstration of the latching mechanism that would later connect the US laboratory module *Destiny* to a central truss for the thirty-year lifetime of the International Space Station. The procedures for this operation were more than forty pages long with hundreds of obscure commands that were not intuitive to an operator, especially this pilot. None of my crewmembers were available to support me in a *two-person* execution of these error-prone procedures, as they were busy with other scheduled events. Thankfully, I had the best flight controllers on the ground, assisting me as a virtual second person. Together, using the radio to communicate, we executed the four-hour procedure without error.

In training, I found some crewmembers occasionally misapplied the two-person rule. When only one crewmember was looking at the checklist while reading it aloud, and the second participating crewmember was performing the specified action separately, effectively only one brain was working on each phase of the operation. To use the two-person rule properly, both crewmembers should view the checklist together, and both crewmembers should observe the preparation, execution, and results of the specified action conducted by the single crewmember.

When used effectively, this two-person buddy system can help operators prevent accidents and save lives. An instance of saving a life might occur after a single climber forgets to connect the fall protection harness before climbing, but a teammate captures the error before they begin their ascent.

One of the better practices to use when following the two-person rule is to "telegraph" actions. This practice affords the second person the opportunity to capture an impending error the first person is about to make.

Example: "Telegraphing" Actions on STS-113
On my final mission, when we conducted operations using the two-person rule, we telegraphed our actions. Before moving any switch or making command inputs to a control device during operations, we indicated our intentions by pausing briefly over the switch or control to allow a second crewmember to verify whether the intended action was correct. Most of the time, the second crewmember merely reported, "I concur," as the first crewmember was about to execute the correct action. Occasionally, the second crewmember captured an impending error by the first crewmember, who may have intended to move the wrong switch or the correct switch to an incorrect position.

When the practice of telegraphing actions becomes automatic between crewmembers, the operating effectiveness of the team improves dramatically. When executed properly, this practice contributes to error-free operations, allowing the team to achieve better performance, with higher-quality results.

5. Identify Trigger Steps (Execution Steps with Immediate Consequences)

A *Trigger Step* is a procedural step that results in an immediate consequence after the action is initiated. Trigger steps are much more important than other steps because something occurs when you "pull the trigger" to initiate the action. Giving yourself the time to ponder exactly what that something might be can save your life.

Many steps in a procedure have no immediate consequence. These steps are preparatory steps intended to configure the system in preparation for the trigger step. In life, most of the actions we take as we walk around during our daily activities have little or no immediate consequences. Most of the time, we don't need to predict the outcome of these actions because nothing bad will happen. But if we learn to identify trigger steps, even in nonhazardous environments, we can prevent bad consequences and save ourselves trouble, inconvenience, or wasted time and effort. In hazardous environments, this technique can save lives.

In the third technique, Follow Procedures (and Rules) *Thoughtfully*, I included a bullet in the meta-level process specifying the operator should predict the intended consequence before taking the action. This helps the operator avoid errors. This ability to identify trigger steps becomes one of the more powerful methods for predicting and preventing accidents and improving performance. For operators who deeply understand operating excellence, trigger steps hold special significance. Here's why.

Photo 49: Telegraphing actions— signaling intent to the second crewmember, before taking action on STS-113 (Photo Credit: NASA)

Example: The Concept of "Trigger Steps"
This first example is a story about the origin of the concept of trigger[18] steps. When the last space flight before our mission rolls to a stop on the runway, we become the prime crew— next up on the launch pad. Training increases in significance. Events in the simulator seem more real. As we approach our launch date, the pressure to perform, along with the scrutiny, becomes more intense. Managers, engineers, instructors, and flight directors are all watching to see how we are operating in our final month before flight. One great advantage for the prime crew is, in addition to watching, these other people are also helping. Flight crews with recent experience perform a support role and provide valuable advice.

---

18   When I created this technique, I referred to these steps as *"execution"* steps. In the space program, we execute the arm and fire commands to the flight control computers. Years later, when I was explaining the technique to Steve Robinson, one of the best drillers in the oil-and-gas industry, he used the term *"trigger"* steps, which is more descriptive.

While we were training in one of our simulations with a large cadre of engineers, flight controllers, and instructors watching, the flight crewmembers made a small error in controlling the Remote Manipulator System (the "arm") and nearly impacted one of two fragile coolant loops of the Space Shuttle. Fortunately, the flight controllers in the Mission Control Center were tied in via telemetry, and provided a quick warning to the crew three seconds before impact.

The need to learn from this small mistake, one month before our flight, was urgent. The instructors explained the potential danger. Had the arm impacted the loop and caused a leak, the failure could have cascaded via a single-point weakness to a total loss of cooling, requiring an emergency deorbit, with much higher risk to the crew.

This is how the error occurred. The arm on the orbiter was similar to a human arm, with a two-degree of freedom shoulder joint, a one-degree of freedom elbow joint, and a multiaxis wrist joint. A similar and slightly more complicated arm is shown on the International Space Station, depicted in photo 50.

Photo 50: The Space Station Remote Manipulator System, or arm, just to the left of the cockpit windows on *Endeavour* during mission STS-113 (Photo Credit: NASA)

Occasionally, depending upon reach limits, the elbow joint had to swing around in a large arc to accommodate a small rotation of the wrist. To keep track of the large movements of the elbow, another crewmember monitored the motion with one of several video cameras mounted in the payload bay. In this minor error, the crewmembers unknowingly used an incorrect camera, so they didn't see the near impact. I knew how easy it is to make a simple mistake. But even a small error in space can be deadly.

Because I was responsible for the safety of my crew on the next flight to leave Earth, I listened intently to the recommendation offered from a crew with recent experience to help us avoid mistakes with potentially severe consequences.

Essentially, they told us, "You have to concentrate, 125 percent, for the whole two weeks up there."

I will always be grateful to the experienced crew for advice. But I couldn't do anything with this recommendation. Humans can't concentrate that intently for such a long time. This really worried me about a month before our scheduled launch. What would I tell my crew? How could I help them (and me) avoid a simple mistake that could quickly become a dangerous event? On its surface, the phrase "to concentrate 125 percent" didn't make sense, but I understood the intent. To the point, though, I knew we could not sustain high-level concentration for more than a few hours, let alone two weeks.

Part of me was surprised, with all my experience in hazardous operations, I never really thought about this problem of how to prevent small mistakes from escalating to disaster. More than surprised, I was increasingly concerned. I had no solution. After discussing this most fundamental challenge with a few other astronauts, I realized they had no acceptable solution either. In a dangerous business, we have to prevent even small mistakes or we might die.

I didn't sleep much that night. I had to face my crew the next day, and I didn't know what to tell them. The more I thought about the issue, the more concerned I became. There is no way, even as an elite crew with the best training on the planet, that we would operate flawlessly for two weeks under high physical and mental stress, with thousands of actions in dangerous operations to construct a space station in low earth orbit. If we couldn't avoid small mistakes, how would we be able to prevent escalation to disaster in our complicated, dangerous mission? It didn't make sense.

About four o'clock in the morning, it suddenly hit me. No human can concentrate at such a high level, continuously. But the trick was, we didn't have to. All we really needed to know was *when* to engage the fifth gear in our brains. We only had to concentrate intensely at certain times. During those times, even the smallest mistakes could not possibly surface without being captured by our best techniques. The rest of the time we could relax and save our mental energy for the can't-fail, must-operate-with-perfection situations.

To anticipate those critical situations, we had to learn to predict certain *trigger* steps during our operations. When we were about to execute a step that had immediate consequence, a step that mattered, it was time to bring all our mental skills, teamwork, and operating excellence to bear on flawless execution.

Most of the time, even in space, there is little consequence for the actions we take. If the arm operator is not commanding the arm to move, it won't hit anything (assuming no failures in the system). If the arm operator *is* commanding the arm to move, but the arm is in places nowhere near the structure, again, it can't hit anything.

In these instances, there is no consequence for taking an action, even by commanding the arm to move in a wrong direction.

In both cases the crewmembers can relax mentally, building up reserves in their cognitive tanks. But now think of a third situation. When the arm operator is about to command a movement of the arm when *any part of the arm* can be close to the structure, this predicted encroachment represents a trigger step. This is the time for all crewmembers participating in the operation to pause, concentrate, and engage all of their knowledge, skill, and attitude for operating excellence to ensure they haven't made a minor mistake that could escalate to disaster. Easier said than done, but, at least, I had a solution. If we could learn to identify trigger steps, we would know when we had to concentrate.

Preparatory steps have little or no consequences. Trigger steps are different because the results can be significant. The best operators develop the skill to identify trigger steps before executing, and they treat these steps differently. Extra vigilance is warranted to ensure the conditions are appropriate and errors have been eliminated before performing these steps, which have greater consequences.

Example: Auxiliary Power Unit Start

In the Space Shuttle, twenty-four switches control the auxiliary power units, which supply power to the three hydraulic systems. Photo 51 shows the control panel on the pilot's side of the cockpit. Five minutes before liftoff, the pilot must properly configure these twenty-four switches to feed the motors with hydrazine, which spins them up to 84,000 rpm.

Photo 51: The pilot's hydraulics panel on a Space Shuttle

No immediate consequence results from introducing a latent error when positioning the initial twenty-one switches of the twenty-four, which simply configure the systems to receive the explosive hydrazine. But the final three switches are different. These are the "Start/Run" switches (indicated by the oval in the photo) for the three independent units. If the pilot has introduced an error in any of the three systems when configuring the initial twenty-one switches, the back end of the vehicle may explode sometime after the pilot moves the final three switches to Start/Run. So these final three switches represent trigger steps in the checklist.

Good pilots use a different mental process to verify all errors have been eliminated after configuring the first twenty-one switches and before moving the final three. Simply repeating the checklist with heightened awareness (*redundant verification*) may be insufficient because the same mistake could be made, even by two crewmembers with similar erroneous thought patterns. A better, more reliable method for eliminating errors is to use *dissimilar verification*. Good pilots execute the checklist carefully the first time. But then, just before moving the final three trigger-step switches, pilots who understand operating excellence will put the checklist down and use a different mental process of looking at the console to identify any asymmetries in the configuration of the first twenty-one switches compared to their mental model of the memorized switch positions. Dissimilar verification has a greater chance of capturing all latent errors.

———

Here is the background for the next example I use to illustrate the importance of identifying trigger steps.

After graduating high school, our eldest daughter, Kelly, told her mother and me she decided to enlist for military service. We were surprised at first, then proud and supportive. She chose the US Army, based on her desired military occupational specialty. After scoring well in boot camp and demonstrating a particularly high aptitude in her chosen field, the Army granted her first choice to become a psychological operations specialist, assigned to US Army Special Operations. I pitied the enemy with her in Psy Ops. The inset picture in photo 52 shows PFC Wetherbee on a successful eighteen-month deployment in and around Baghdad, Iraq.

A prerequisite for being designated a psychological operations specialist was to be qualified as an Airborne soldier. The US Army Airborne School in Fort Benning, Georgia, is one of the training programs I admire in the Army, from the perspective of an outside observer. Airborne training culminates with each private executing five successful parachute landings, the final one at night, from a C-130 aircraft flying at low altitude.

But, here was the best part for the parents of their firstborn daughter, who was going off to war in service of our country. We were invited to visit Fort Benning and watch the two-day exercise. I have had the privilege of seeing some amazing things in my life. Witnessing the final night qualification drop for the young privates in the US Army was an incredibly moving experience.

Initially, from the edge of the drop zone, we didn't sense much in the dark quiet night. Soon we heard the barely audible hum of the C-130 on its approach. After multiple passes over the drop zone, we faintly saw the fearsome outlines of scores of dark parachutes in the night sky, dropping silently. About twenty minutes later, we began to see from a distance all the troops marching single file up from the drop zone, gradually becoming more and more illuminated by the bright spotlights near the stands where the parents were seated, listening to our national anthem playing over the loudspeaker. This was the US Army, doing it right.

The following is an example, from Airborne training, that illustrates the benefit of developing the skill to identify trigger steps, in the pursuit of operating excellence.

Photo 52: Drop Tower for Parachute Descent Landing Training, Airborne School, Fort Benning, Georgia. Inset picture: PFC Kelly D. Wetherbee, Airborne, psychological operations specialist, on deployment in Iraq

Example: Drop Tower for Parachute Descent Landing Training, Airborne School, Fort Benning, Georgia

Before jumping out of a perfectly good airplane for the first time, privates in the Army's Airborne School receive prejump training on the parachute drop tower (photo 52, shown to the left). This is a buildup exercise to give the students, and the instructors, confidence their first jump from the airplane will be successful.

The drop tower is 250 feet tall. The Airborne candidate is harnessed to an open parachute and hoisted to the top of the tower, as shown in photo 53. At this point, the student goes through the checklist memorized for execution before being dropped from the tower. Each item is important and must be conducted in the proper sequence. Forming consistent routines and good operating habits saves lives in hazardous occupations. The instructor on the ground is also going through a checklist to verify the student has performed the routine correctly and is in the proper body position for landing.

This was the time for the student to take one last breath and be ready for the drop. This was also the time for me to pick up my camera, as I suddenly noticed something different. Depending upon the next decision the sergeant would make, the drop could be worth capturing on my camera. From my vantage point in a parking lot at the edge of the field, I had seen a few of the students drop successfully earlier. But, because of what I was seeing before this release, I thought, *Whoa, this one's going to be interesting.* If you look closely at the top of the tower in photo 53, you can see the same thing. Based on the direction of the windsock, which way will the parachute drift if the student who is lower in the photo is released?

Sure enough, that student was released first. You can see the results of the unsuccessful drop in photos 54–56. From more than one hundred yards away, I could hear screaming as the parachute drifted into the tower and became entangled. I could see the student kicking his legs wildly, and I thought, *Dude, stop kicking. You're going to rip your 'chute and plummet to your death!* In retrospect, this outcome may not have been likely, as the parachute has strong nylon cords and a rip-stop configuration. But, I doubt he was making his situation any better by kicking wildly, and I don't know the weakness of that part of the tower or the fragility of the parachute entanglement.

What I saw next was truly amazing. The initial part of the rescue operation to secure the student and prevent him from falling took no more than ten seconds. (You can see the method of the rescue operation in photo 57.) In typical military preparedness fashion, all the other students were standing by, before and during every

Photo 53: Drop tower, with two students preparing to be released

Photo 54: Drop tower—drop sequence

Photo 55: Drop tower— drop sequence

drop sequence, manning a long line rather than simply standing around waiting for their turn. At the instant the parachute hung up on the tower, the instructor hooked himself to the line. On command, the whole class immediately ran at full speed away from the tower. As they removed slack in the line, the instructor was lifted straight up from the ground and, with the skill only to be achieved through practice, came to a stop right next to the student and pulled him to safety.

So what's the point? As you read the story, I hope you identified the trigger step that the instructor could have identified, which represented the last time to prevent

Photo 56: Drop tower—drop sequence

Photo 57: Drop tower—rescue operation

this incident. When the students were hanging in their harnesses, and they were going through their checklists to verify the configuration of their parachute, harness, body position, and readiness to be released, no immediate consequence would have occurred for failing to complete any of the preceding steps. At that moment, they were only hanging there. When the instructor on the ground was going through his process to verify a student had completed the checklist and was ready to be dropped, again there would be no immediate consequence had the instructor failed to verify the student's readiness.

But suddenly, the instructor was moments away from the trigger step (sending the command) to release the student and begin the drop sequence. Immediately before the trigger was pulled, the instructor should have taken the time to step back and think, *I'm about to do something that will have consequence; have I checked everything one last time? Have I forgotten anything? Oh yeah, I forgot to check the wind. Better not pull the trigger just yet.*

Developing the skill to identify trigger steps before they occur is the biggest challenge in this technique. What you do after identifying the trigger step to prevent an adverse consequence is much easier. I'll get to that shortly. But for now, I include a few more examples to illustrate the concept.

The identification of trigger steps helps in everyday life, as well. Forgetting your house keys only has a consequence after you close the locked door behind you. The act of closing the door has greater significance than the footsteps taken earlier when distracted by thinking of your upcoming meeting. If you are conditioned to predict the impending consequences of trigger steps, you may think about your keys just *before* the door closes and locks, rather than just *after*.

Example: Robbery in a Hotel Room, Scene One

For several years, I had been teaching the concept of identifying trigger steps. Usually, I told the story of a test pilot friend who was staying in a hotel near Fort Walton Beach, Florida, during a deployment for the test and evaluation of an aircraft system. One afternoon, he made plans for a friend to meet him at his hotel room at 6:00 p.m. to depart for dinner. At 5:45 p.m., he heard a knock on his hotel door. When he opened it, expecting his friend, he found himself staring into the barrel of a handgun, held by a burglar who forced his way in to conduct a robbery.

The thief duct taped my friend's hands behind his back, put him face down on the bed, and stole $400 in cash. Just before the criminal fled, he told my friend not to move for thirty minutes, or he would come back and shoot him. The story has an acceptable ending, at least for my friend. Thirty-one minutes later, he got up from the bed, unhurt, and walked to the front desk of the hotel. He told the manager about the robbery and advised the manager to expect no further government employees would be staying at the hotel unless he was reimbursed $400. The manager acquiesced.

The story illustrates the point that hearing a knock and walking to a hotel door is not a trigger step. There is no consequence for approaching the door. Putting your hand on the handle of the door with the intent to open still has no consequence. But suddenly, once you rotate the handle, you have executed a trigger step because you might end up staring down the barrel on the wrong side of a trigger. Better to identify the step, anticipate the consequence, and look cautiously through the peephole before executing the trigger step of turning the handle.

Example: Hotel Room Door, Scene Two

Somehow, the teacher didn't learn. Years later, I was staying in a hotel in Rock Springs, Wyoming, with twenty colleagues. One afternoon I heard a knock at my hotel door, shown in Photo 58. After correctly identifying the trigger step, I carefully looked through the peephole to see who was knocking. Suddenly, my finely-tuned, highly-skilled, appropriately-cautious thought process was easily hijacked by an automatic subroutine in my brain. After I was surprised to see the peephole was cloudy, I instinctively opened the door. In barged a large, muscular man who said in a booming voice, "Where's Mike?" The scary man, who I later found out was a Special Agent at a Secret Service convention, suddenly realized his mistake and said, "Sorry, wrong room." I suddenly realized my mistake, a flood of adrenaline later.

Even though I had been teaching the technique for years, my flawed instincts took over when the deviation occurred. After correctly identifying the trigger step, I failed to take correct action before pulling the trigger. When I couldn't see through

Photo 58: Hotel door (Rock Springs, Wyoming)

the peephole, the hazard was unchanged. I should have called "Who's there?" through the closed door to avoid the adverse consequence.

―――――

Example: A Useful ELCC Checklist
In airplanes, good pilots always think about their actions in terms of the follow-on possibilities. Before pushing the ignition button in the cockpit to start their engines, the best pilots prevent damage to equipment or people by reviewing one last time that they have set the proper configuration, enabled electrical power and fuel flow, and that the external area is clear.

Even after diligently completing their takeoff checklist, pilots know they might have made a mistake that could lead to an accident during takeoff. So they mentally check critical items one last time just before entering the duty runway (Photo 59).

Pilots understand the takeoff event is a trigger step with potentially catastrophic consequences. As they cross the threshold to the runway, good pilots prevent their

Photo 59: Just prior to takeoff in a T-38 during Space Flight Readiness Training before mission STS-86, Kennedy Space Center, Florida (Photo Credit: NASA)

next potential accident by re-checking the status of the only items that can kill them in the next minute: their *Engine(s), Lights (warning and caution panel), Configuration, and Controls*. Quickly re-verifying four items, *ELCC*, gives them simple assurance they will avoid potentially fatal consequences. All critical checklist items that might have been overlooked have now been captured, immediately prior to the trigger step of takeoff.

So far, I have written mostly about identifying trigger steps before executing them. Here are the detailed, systematic thought process and actions to take after identifying the trigger step. Immediately prior to executing the step, elite operators think of three operating situations to reduce exposure to consequences. To help me remember these *three Ls*, I think of the concepts of past, present, and future:

A. *Line-up* (Past). Has my configuration been *Lined-up* properly? That is, have all the devices upstream and downstream in this line—fuel line, steam line, propellant line, hydraulic line, or electrical line—been set in the proper configuration prior to taking this next action? Devices include any type of software and hardware configured in the system, such as valves, pumps, blocks, electricity, flow, pressure, lines, diverters, filters, signals, switches, controls, engines, etc. (This is the verification I failed to perform during the emergency we experienced on STS-32 when I connected a hose to a pressurized waste water tank.)

B. *Line-of-Fire* (Present). Am I or any other personnel in the *Line-of-Fire*? Will any person, equipment, or system be damaged as an immediate result of triggering this action and releasing energy?

C. *List of Results* (Future). What will occur after I take this action? What is *expected* to occur? What will I see, hear, or sense if the expected result occurs? More importantly, what might happen that is *unexpected;* in other words, what should not happen? What will I sense if something happens that shouldn't be happening? What actions will I take if the unexpected occurs? Do I run, reverse the action, call for help, get a fire extinguisher, or sound an alarm? The best time to think about these actions is not after the unexpected happens, when the surprise can kill. The best time to prepare is before executing the trigger step.

When mentally performing the three Ls of Line-up, Line-of-fire, and List of results, the best operators use dissimilar verification methods to make sure they and their systems are ready for the step to be executed. Dissimilar verification, which I will present later, provides the greatest opportunity for capturing latent errors.

Example: Line-up—Verification before Opening Valves

Assume you are an operator working offshore on an oil platform, as shown in photo 60. If you were tasked with opening a valve, how do you know the pipeline is properly configured upstream and downstream from your valve? Do you trust other operators to have conducted their procedures correctly? The safe course of action, consistent with the philosophy of operating excellence, is to verify your line-up configuration prior to executing the trigger step.

Photo 60: Valves and pipes on an oil platform

Example: Line-of-Fire—Injured by a Drain Pipe

Offshore drilling platforms are fabrication wonders that represent amazing feats of engineering design. A movable drilling rig can be used on a floating platform at sea to position a drill pipe precisely, within inches of the desired location above the well. The movable rig is on giant tracks oriented to translate in north-south and east-west directions relative to the platform. In this example, one of the rigs had a design problem. Each time the rig was moved, a three-inch diameter drainpipe on the rig scraped along a handrail that was part of the fixed platform. The temporary solution was to cinch a tension strap around the drainpipe and bend it slightly away from the rail before moving the rig.

After the move was complete, an operator was tasked with releasing the ratchet on the tension strap. Before the job was scheduled, a job-safety-analysis was conducted, and a particular hazard was identified and described. According to the analysis, when the operator releases the ratchet mechanism, the removal of tension on the drainpipe releases the stored energy in the bent drainpipe causing the pipe to spring back toward the handrail. To protect the operator from injury, the control mechanism specified was to "use good hand-and-body position" prior to releasing tension. The operator initialed the written job-safety-analysis, indicating he understood this control, to not place his hand between the drainpipe and the handrail.

The next day, the operator went to the jobsite and executed the task. He needed to lean over the handrail to reach the release mechanism with his left hand. The automatic processing in his brain placed his right hand on the handrail. When he released tension, as predicted in the analysis, the drainpipe snapped back and struck a finger on his hand holding the rail.

I used the previous example is to illustrate that some control mechanisms are ineffective when they rely on an action the brain normally conducts automatically. As leaders and safety professionals, we must figure out ways to help our workforce prevent injuries. Trigger steps can be identified in written procedures. Operators can be trained to develop the skill of pausing before trigger steps are executed to verify safe conditions are in place. Front-line supervisors can observe and verify the operators are developing the skill to automatically check Line-up, Line-of-fire, and List of results, one final time before executing any trigger step. This technique gives the operator one last chance to prevent injury by replacing conventional automatic processing in the brain with new automatic skills purposefully trained.

Example: List of Results—Expected or Unexpected Outcomes?
The following picture shows a liquid hydrogen line (see photo 61). In which direction do you expect the flow to go if you take an action to reconfigure a valve on this line? What will you do if something unexpected happens? When do you think about your response, before or after you open the valve?

Before taking action to change the configuration of the line, predict the expected and unexpected results, and know what you will do if the unexpected happens.

6. Perform Verification

Successfully accomplishing missions requires integration of many large-scale processes, medium-scale activities, and small-scale individual actions. Because of the interrelated and complex nature of the systems and processes, any small malfunctions, failures, or errors that have undetected adverse consequences can cascade quickly to major catastrophes.

Photo 61: Liquid hydrogen line. Before taking action, predict the expected and unexpected consequences, and know what to do if the unexpected occurs.

Verification is about the search for truth. If an operator issues a command to open a remote valve, the operator must ascertain whether the valve opened properly, as expected, or not. To prevent accidents, operators need to be able to identify all adverse consequences and take corrective actions before any adverse consequence can escalate or compound into a larger problem.

When using this technique, *Perform Verification*, operators will be able to identify malfunctions, failures, errors, unexpected results, and adverse consequences. Three grades of verification provide increasing amounts of confidence in the determination of truth.

A. Simple Verification

This is the least complicated method of verifying the current state of a system. Cross-checking is an example of a verification method. An operator can simply review the procedures after completing a group of steps to verify they were executed correctly.

Example: Verification of Survival Suit Pressure Integrity

On the Space Shuttle, the potential hazard of cabin depressurization was greatest during two phases of flight. For the first few minutes of powered ascent from the launch pad, 7.7 million pounds of thrust are pushing 4.5 million pounds of hardware, propellant, and people through the thick part of the atmosphere at high dynamic pressure. All the combusting propellant is converted to fire, smoke, and kinetic energy, which, after eight minutes and twenty-three seconds, allows us to float lazily at five miles per second of orbital velocity. Two weeks later, that same amount of energy must be removed by slamming into the atmosphere at hypersonic velocity, which creates a trailing plume of fire twenty-five miles long that sways back and forth behind the runaway freight train that is the Space Shuttle orbiter.

Because of the risk of cabin depressurization during these two violent phases of otherwise peaceful flight, astronauts wear pressure suits for ascent and entry. The bulky uncomfortable suits become our friends, since they have the potential to save our lives. The suit technicians, who verify the pressure integrity of our suits before flight, become our good friends.

Photo 62 shows a suit technician verifying the integrity of my pressure suit before the launch of STS-86. By the way, because of the importance of this task, I verified what he verified.

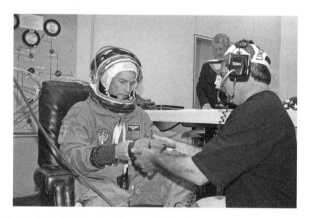

Photo 62: Suit technician verifies the integrity of survival suit pressure integrity, STS-86 (Photo Credit: NASA).

Example: Dropped Loads in a Trucking Company[19]

A trucking company in the US realized they were experiencing occasional dropped loads with their tractor-trailer rigs on highways. Their analysis showed the inadvertent separations of the trailers from the tractors were occurring with experienced drivers and not with the newer drivers. When the company investigators interviewed some of their experienced drivers, an interesting commonality was discovered. The drivers reported they could hear a distinctive sound when they were connecting the hitch on the tractor to the trailer (similar to photo 63). This was their auditory technique for verifying a solid connection with a good latch before departing with the load. The less-experienced drivers reported a different— and more reliable—method of verification.

---

19   I heard this story from Walt Schumacher, a colleague and highly-skilled manager in labor and
     employee relations.

After the less-experienced drivers were told about the technique of hearing the latch drop home, they each responded with statements similar to, "Oh no, I can't hear it; I have to get out and look to see if the latch is secure." The less-experienced drivers performed their visual verification even if it was raining. The company wisely promoted a new mantra, describing their approved process of verification: "It's not a good hook unless you get out and look."

Photo 63: Hitch mechanism on tractor-trailer. Visual verification of good hook engagement is required before departing with load.

B. Redundant Verification

In the control room of a refinery and on aerospace vehicles, some systems have redundant sensors to indicate the state of parameters, such as, the position of a valve, the speed of a motor, or the pressure in a tank. The output of the sensors may be sent to two different types of displays (for example, a hard-wired mechanical gauge and a video monitor). Verifying the output of the sensors on two different displays is a redundant method for cross-checking the value of the parameter. Using two crewmembers to complete a procedure can be considered a redundant method of verifying, or cross-checking, the accuracy of a process.

Example: Verifying Gear Down
The Space Shuttle had a unique hazard (among hundreds of others) that exposed the crew to the potential of fatality. Because of the excessive amount of aerodynamic drag on the vehicle during final approach, the deployment of the landing gear had to be delayed until seconds before touchdown (see photo 64). With no engines available to abort the landing, the crew had little time to troubleshoot any problems with the landing gear and take corrective actions.

Immediately after issuing the command to deploy the landing gear, the pilot had only seconds to conduct a verification process by looking down in the cockpit to see proper indications of successful deployment. After verifying, the pilot verbally reported quickly on the intercom, "Gear down," while the mission specialist confirmed redundantly the hazard had been controlled and no further actions were required. Carrying the *Redundant Verification* process even further, because we really didn't like single-point failures, some commanders issued a simultaneous redundant command to deploy the gear. The mission specialist could verify visually both the commander and the pilot pushed their buttons to deploy the landing gear. Multiple levels of redundant verification occurred within seconds to prevent the next potential accident being caused by numerous unknown, but anticipated, failures.

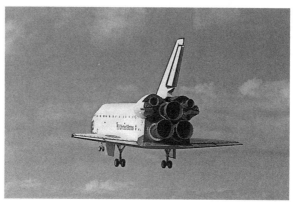

Photo 64: Seconds before touchdown, pilot has verified the successful deployment of the landing gear on STS-113 (Photo Credit: NASA).

Example: Tether Protocol for Space Walks on ISS

Notice the grandeur and beauty of the scene depicted in photo 65. The majesty of it all suddenly becomes trivial and irrelevant if you find yourself inadvertently untethered and drifting away from your safe harbor into the blackness of space. Know that making such a simple mistake is easy. For this reason, we spend an enormous effort to redundantly verify that every spacewalking crewmember follows rigorous, double-tethering protocol at all times in space. Visually and verbally checking and double-checking, by crewmembers inside and out, gives us the confidence to know a simple mistake will be trapped quickly, so we can continue to appreciate the grandeur and beauty of the scene.

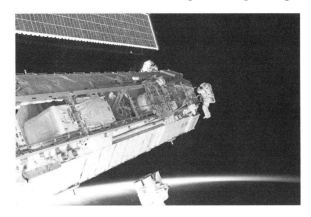

Photo 65: John Herrington and Mike Lopez-Alegria use redundant verification of proper tethering during spacewalks on the International Space Station, STS-113 (Photo Credit: NASA).

Example: Verification of Engine Performance with Redundant Computers
The power of redundant verification is carried over into the realm of systems and computers, as well. Photo 66 shows the displays in the cockpit of the Space Shuttle simulator. Redundant computers used processing logic to calculate redundant information from redundant sensors on vehicle systems for display on redundant monitors, to be interpreted by redundant crewmembers. Any single link in this information chain could, and sometimes did, fail. Redundant verification helped us sort through the various possible failures to identify the culprit and take remedial action.

Photo 66: Front cockpit layout, one minute and forty-one seconds after liftoff in the Space Shuttle simulator. Displays show slightly different values from the main engines' sensors, calculated from redundant flight computers.

In the previous example, I alluded (redundantly) to a limitation in the concept of redundant verification. Are two redundant computers calculating thrust vector control for the propulsion system better than one computer? Yes, but only to a point. The two computers, though operating independently, might have the same erroneous processing error typed into their programming logic. Two redundant crewmembers likely had the same training, read the same flight manuals, and may have been shaped and molded throughout their military careers as test pilots to think the same way, even if erroneously. Redundant verification doesn't work if two systems or crewmembers are identically incorrect. This illustrates the power of diversity of thinking. It also leads us to the third and most effective method of verification.

C. Dissimilar Verification

By far, the best method of establishing truth is *Dissimilar Verification*. When a valve in a gas line is opened, redundant position sensors may indicate the actual position of the valve. But these redundant sensors may have the same erroneous bias from a faulty signal converter, or other single-point failure mode, depending upon the design of the system. The redundancy is lost if the single failure takes out both sources of information. But if the operator has access to a different kind of sensor than the valve position indicator, such as the pressure sensor downstream from the valve, the accuracy of the *position* sensor can be verified with the dissimilar *pressure* sensor. This provides greater confidence the system is working correctly and the valve truly is open.

Similarly, after an operator has completed a procedure, a simple review of the procedure may lead to the same erroneous mental process that led to misreading or omitting a step. When the operator uses a dissimilar method of verification, such as looking for expected symmetry in the switch positions, an error may be identified before an adverse consequence occurs.

An easy method of dissimilar verification can be used after mentally adding a string of numbers. Verifying the result by adding the numbers a second time but in reverse order may prevent the same carry-over error from being made twice.

Example: Verifying the Switch Position Checklist using Dissimilar Verification
Strapped in tightly to a rocket on a launch pad is the perfect time to appreciate the privilege of being selected to fly for our nation's space program. It is also the perfect time to not make a mistake. I wanted to identify any pre-launch errors in the positioning of critical switches when safely on the ground, rather than unsafely not on the ground immediately prior to an in-flight disaster.

In this can't-fail situation, I used a process of dissimilar verification to ensure hundreds of switches and circuit breakers were in their correct position prior to liftoff for each of my six flights. Photo 67 shows me in the commander's seat, using the checklist to verify the cockpit switch positions before launch on STS-113.

After I verified every switch position the first time by methodically using the checklist, which might have contained an unidentified error, I used two additional dissimilar verification methods, with two different thought processes. To verify the proper positions a second time, I simply looked for mismatches in the visual pattern of correct switch positions I had previously memorized. The third time, I used an even more reliable method. Thinking more deeply about how each system operated, I verified each switch was in the correct position to properly configure every component allowing the system to function as intended.

Photo 67: Using a checklist to verify cockpit switch positions before launch on STS-113. Dissimilar verification is used to ensure correct configuration of switches in this critical ("can't fail") situation. (Photo Credit: NASA)

Systems on airplanes are highly reliable. Even so, in 7,000 hours of flight time I have experienced hundreds of system failures. To operate safely with system failures, the pilot's first task is to identify which system failed. In today's complex systems that are highly integrated and controlled automatically by computers, identifying which system failed and how it failed is often challenging. Dissimilar verification is useful in identifying failures.

Example: Landing Configuration in Naval Aviation
Here's an example to illustrate how dissimilar verification is used in flight operations. Most pilots rely on the airspeed indication, provided by the pitot static system, for flight control during all phases of flight, including landing on a runway. Redundancy

is built into the pitot static system to reduce the likelihood of a complete failure of airspeed indication. Even with redundancy, meteorological conditions at high-altitude created icing in the pitot static system that caused a failure in the airspeed system on Air France Flight 447 in May 2009. The flight ended in catastrophe as the airplane crashed into the Atlantic Ocean.

Some pilots are trained to use dissimilar systems when they experience the complete failure of one system. Without proper airspeed indication, pilots can still control the airplane safely, using their knowledge of the proper flight path angle of attack and engine power setting required at any particular altitude. The tolerances for acceptable flight path angle of attack are quite narrow when landing aboard an aircraft carrier. Naval aviators are trained to use the angle of attack indication during arrested landings at sea. The angle of attack system and the airspeed system are independent. Either can be used for landing. Both are used in a method of dissimilar verification to ensure proper flight control.

Photo 68 shows the view through the head-up display in a US Navy A-7 Corsair. With airspeed and angle of attack displayed simultaneously, the pilot can use dissimilar verification quickly and easily before every landing to discern failures in either system.

Photo 68: View through head-up display just prior to landing in the A-7 Corsair. Naval aviators are trained to confirm proper landing configuration using a method of dissimilar verification (cross-checking angle of attack with airspeed).

Example: Dissimilar Verification of Fuel Quantity in the A-7
This final example of dissimilar verification is presented to suggest the method can be used in different industries, or in everyday life, to prevent accidents.

Pilots in airplanes and drivers in automobiles use a fuel quantity gauge to know when it is time to refuel. But, what if the gauge experiences a failure and sticks in one position and the pilot or driver unexpectedly runs out of gas? In a car, the driver can walk to the nearest gas station. Same thing in an airplane, but only if the pilot was able to dead stick to a successful ditching and then walk to the gas station. To avoid hearing the engines suddenly go silent in flight, pilots continually use a dissimilar verification method to assess the accuracy of their fuel quantity gauge by integrating fuel flow rate over time (see photo 69). The fuel flow sensor and indicator are independent from the fuel quantity sensor and indicator, so errors can be identified readily. For planning purposes on long trips, car drivers know they can't drive farther than a certain range on one tank of gas, regardless of unidentified inaccuracies in their fuel quantity gauge.

Photo 69: Front cockpit layout of the A-7 Corsair.
Oval indicates fuel quantity and fuel flow gauges.

Numerous accidents in various hazardous industries have occurred after operators believed static or erroneous quantity indications in tanks being loaded or filled with explosive substances. Had the operators been trained to use dissimilar verification methods by comparing quantity with flow rate integrated over time, they might have been able to resolve the errors to prevent the accidents.

7. Protect Equipment and Systems

This technique is intended to help operators prevent failures and extend the useful life of their equipment and systems by taking care of the equipment and systems they control.

Good operators always think about risk and always understand the vulnerabilities of the systems they control. Good operators protect their systems by using their equipment as designed, within operating limits, and in optimal sequences.

Hardware responds deterministically. It doesn't just fail. A specific component will fail for a particular reason. If we knew every reason, we could prevent every failure. Unfortunately, operators are not omniscient, and some of the reasons are complex or obscure. But, even with incomplete knowledge of how specific failures occur, operators can minimize the chances of provoking component failures by treating the systems well and operating the equipment and systems within design envelopes, operating limits, and known guidelines.

As an operator, I know the vulnerabilities of my systems, and I must not push their redline limits. I must *Protect Equipment and Systems* to prevent failures. If I take care of my hardware, it will take care of me.

Example: Space Shuttle Challenger *Accident*

Here is the tragic example of the *Challenger* disaster to illustrate that hardware does what it is allowed to do and fails for particular reasons.

The white solid rocket boosters (SRBs) on each side of the orange external tank were fabricated in four cylindrical segments, twelve feet in diameter. The segments were connected at their circumference with a structure called a field joint that formed the seam between each segment.

Managers decided to launch the *Challenger* on January 28, 1986, in violation of a launch rule that specified a lower limit for ambient air temperature at the launch pad. The rationale for having a temperature limit in the launch rule was to protect the hardware by ensuring two rubbery O-rings in each field joint would be warm enough, and therefore compliant enough, to seal a gap in the field joint and prevent internal combustion gas from escaping through the seam after ignition.

As soon as the SRBs were ignited in ambient temperature below the specified limit, hot internal combustion gas began to escape past the cold, noncompliant O-ring seals in one of the field joints. The hot gas blow-torched the attachment mechanism, similar to a strut, connecting the SRB to the external tank. After seventy-three seconds of superheated gas impingement, the attachment mechanism melted and failed, allowing the SRB to rotate into the external tank, causing structural failure of the external tank, catastrophic deflagration of the vehicle, and death of the crew.

After seeing photographs of the bright, hot gas impinging on the attachment mechanism, some people speculated we were only unlucky at NASA. Had the hot gas escaped anywhere else around the thirty-eight-foot circumference of that particular field joint, or from any other location in either of the other two field joints in the SRB, the attachment mechanism wouldn't have failed and the vehicle might have survived one more minute until normal SRB burnout. The sympathetic people surmised it was terrible luck, which resulted in the field joint leaking in the only location that could have caused structural failure.

These people drew an incorrect conclusion. Hardware doesn't care about luck. The structure that melted was the attachment mechanism between the SRB and the external tank. By design, that mechanism was required to transmit a huge load from the SRB, which was generating one million pounds of thrust, to drag the massive external tank and orbiter into space. That very attachment mechanism caused the greatest amount of deformation—resulting in the largest opening—in the field joint at that particular location simply because that was where the mechanism was attached to the SRB. (On previous launches, the O-rings were warm enough to fill the same large gap in the seam.)

The hardware did what it had to do because it was designed that way. The SRB was dragging the external tank, and the external tank resisted by deforming the SRB at its seam. When the hot gas escaped through the largest opening, it necessarily impinged on the attachment mechanism—which was causing all the stress that opened the field joint in the first place.

The point of this tragic example is that once the managers decided to violate the launch rule by launching in cold temperature, the hardware responded deterministically in the only way it could, by failing catastrophically exactly in the way it did. Neither blind unfortunate luck nor rigorous engineering probabilities had anything to do with this disaster. Once the managers made bad decisions to violate the launch rule, which was designed to *protect the equipment and systems,* the hardware was doomed to fail, with a certain probability of 1.0.

Example: Blowout Preventer in Drilling Operations

Here is an example from the oil-and-gas industry I use to illustrate how good operators can prevent failures by protecting their systems and operating their equipment within proper design envelopes. During drilling operations at sea, a massive structure, known as a blowout preventer, which is the size of a house, sits on the sea floor surrounding the drill pipe that extends miles into the earth down to the hydrocarbon reservoir.

This blowout preventer contains several large devices, which drilling operators can actuate remotely from the surface of the sea. These devices are intended to block the different flow paths from the underground reservoir to the surface. With the paths blocked, the hydrocarbons are isolated and cannot rise uncontrolled in a classic blowout. Each device has different intended functions and different vulnerabilities to adverse conditions.

The good drillers understand the functions and vulnerabilities. In a suspected blowout situation, these skilled drillers actuate the devices in a particular sequence to allow the function of each device to protect the vulnerabilities of the next device in the isolation sequence.

For readers who understand the drilling terminology, here is the sequence the best drillers use to protect their equipment and systems. When uncontrolled flow of hydrocarbons is sensed, the driller initially actuates the *variable bore rams,* which are intended to isolate the flow up the annulus around the drill pipe; the driller still has time to assess the situation. If the hydrocarbon "turns the corner" from the now-blocked annulus and is coming up the drill pipe, the risk to personnel and equipment is increasing significantly. Next, the driller actuates the *casing shears* to cut through the drill pipe. The driller intends this action to diffuse the raging hydrocarbon flow, which will try to erode a vulnerable seal in the final component, the *blind shear ram.* The blind shear ram is actuated last. This is the only device that has the capability to isolate the flow of hydrocarbons completely. But, the flow is isolated only if the vulnerable seal doesn't fail. The entire sequence skilled drillers use is designed to protect this final, vulnerable, "can't-fail" seal in a last-ditch effort to terminate the blowout. Protect the seal in the blind shear ram so it can protect the crew.

Example: Using the Drag Parachute and Light Braking During Space Shuttle Rollout
In the section describing the technique Develop and Maintain Risk Awareness and its subtechnique Search for Vulnerabilities, I presented an example of a particular weakness in the landing and rollout system of the Space Shuttle. Little margin existed between the braking energy needed to stop the Space Shuttle after landing and the maximum braking energy capacity of the beryllium brakes on the four main wheels. To reduce the demand on the brakes during rollout, a drag parachute was retrofitted into the design after forty-six flights. The parachute is shown, just prior to opening fully, in photo 70.

Photo 70: Immediately after nose wheel touchdown and drag chute deploy, STS-113 (Photo Credit: NASA)

Because of this vulnerability in the brakes and my desire to protect equipment, on all five of my landings in the Space Shuttle, I maximized the advantages of aerodynamic drag and rolling friction and used reduced braking to roll gently to a stop. The idea was to protect the vulnerable and valuable hardware, and allow the hardware to protect my vulnerable and priceless crew.

In using this technique, I realized two benefits—one intended and the other unintended. First, I did not blow any tires, and I maintained directional control and rolled safely to a stop on the runway. I had not predicted the second benefit. After we egressed the vehicle and were performing our ceremonial postflight inspection (see photo 71), I had an opportunity to meet the dedicated workers who processed the vehicle between spaceflights.

The landing gear maintenance crew introduced themselves to me after two of my five landings in the Space Shuttle. They shook my hand and thanked me for limiting my use of the braking system. The energy absorbed by the system was low enough they did not have to replace the brakes, as they did after most of the other landings in the Space Shuttle program.

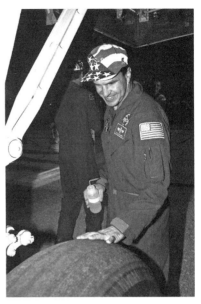

Photo 71: Pilot Mike Bloomfield during ceremonial postflight walk-around of *Atlantis* after STS-86 (Photo Credit: NASA)

Protect your equipment, and it will protect you. You and it will last longer.

Operators must be careful not to limit or reduce the safety margins unintentionally by modifying operations with an insufficient analysis of the possible consequences. Specifically, as I change my way of operating to protect my equipment from one vulnerability, I may be exposing my equipment to another vulnerability. So, before I modify normal operations to protect my equipment, I consider two other techniques, which I will explain later: *Expect Failures* and *Reduce Exposure to Hazards*.

To appreciate how these techniques are related, consider the previous example where I used reduced braking during rollout on the runway. As I was limiting my use of the brakes, I was losing precious runway length in front of me. This means I was reducing the amount of time to react if we had an independent and latent failure in our braking system. Had such a separate failure occurred, I would want to know about that failure earlier, while I still had options to control the hazard, rather finding out too late and rolling off the runway. So I used the normal deceleration profile early in the rollout to verify the braking system was working properly before I relaxed my braking command to limit the energy into the system.

Protect your equipment. Don't exceed redline limits. But, in a dangerous environment, as you reduce the demands on your equipment in your effort to protect it and prolong its usefulness, make sure the operational changes you make do not inadvertently expose your systems or crew to other hazards.

Example: "OK, Atlantis, I'll Take Care of You . . ."

I learned several valuable operating techniques from Dan Brandenstein, my commander on my first flight into space. I emulated many of his practices on the five missions I commanded. One of his habits was to pause briefly at the base of the launch pad, just before boarding the world's slowest elevator to the top of the stack. Dan looked up at the magnificent Space Shuttle system, to which he was about entrust his life. He never told me what he was thinking in those brief moments before he got down to business, but I picked up on his tradition.

Photo 72 shows a similar scene on the day before the launch of STS-86. Just before ingressing *Atlantis* for that mission, I had the same view (without the flag). I reserved that moment for a rare emotional connection with my vehicle, in a dangerous operation that normally requires suppression of emotion. I looked up at our spaceship, which seemed to be alive as it creaked and hissed with millions of pounds of cryogenic propellant trying to boil off, and thought, *OK, Atlantis, it's you and me now; I'll take care of you if you take care of me.*

Photo 72: STS-86 on the launch pad (Photo Credit: NASA)

Example: Accelerating from an Intersection after the Traffic Light Turns Green
How often do you notice other drivers accelerating quickly as soon as the light turns green, only to see they must stop fifteen seconds later at the next intersection, which has a red light (see photo 73)? If you want to protect the equipment and systems in your car, both the engine and the brakes, don't execute jackrabbit starts or screeching stops.

Photo 73: Accelerating quickly from intersection when light turns green, only to stop fifteen seconds later at the next intersection with a red traffic light

Example: Two Spinning Grinders
Here are two similar examples of spinning grinders being operated in a way that caused injuries in a machine shop. In the first example, a large grinder was in a fixed position over a table with a sliding platform used to mount objects intended for grinding, sanding or polishing. When the rotary wheel of the grinder was spinning at high rpm, a red indicator light was illuminated next to the start and stop buttons.

Upon completion of a sanding task, an operator pushed the stop button to deactivate the rotary wheel, and he turned away from the machine. As designed, the red indicator light was extinguished when electrical power was removed. The operator's assistant decided to help by wiping the sand residue from the platform under the rotary wheel. The assistant didn't notice the wheel was still spinning at high rpm with its angular inertia, and there were no indications the machine had recently been powered. As he ran his hand along the platform in a wiping action, the spinning wheel chewed through his glove and injured his hand. Without the design feature of a light to indicate residual motion of the wheel, even when unpowered, or the safety feature of a manual guard, the operator should Protect Equipment and Systems—*and teammates*—by not turning away until the wheel has fully stopped.

In the second example, a hand sander can be seen in photo 74. An operator placed the sander on a table before the wheel had stopped spinning. As the sander jumped, the operator instinctively reached for it. Unfortunately, with his quick reaction, he injured his hand against the side of the table. Again, without the design feature of a guard, the operator should Protect Equipment and Systems by not releasing the sander until all of the mechanical or electrical energy has been dissipated.

Photo 74: Spinning sander. Operator released the sander before it stopped spinning.

8. Expect Failures (System and Human)

In all the years I have spent working in operations, I have noticed failures occur. Equipment, systems, and humans all fail. Though the failures may occur at any time, I have benefited from anticipating the worst failures will occur at the worst times. There seems to be some validity in this pessimistic concept. Failures occur more often when equipment, systems, and humans are stressed. Inconveniently, the greatest stress occurs during activation or transition in operating modes. So at the exact time when the new or changed operating mode is needed to conduct an operation or prevent an accident, failures are more likely to occur.

In using this technique, *Expect Failures*, operators anticipate corrective actions or alternate methods to conduct operations safely and effectively should a failure occur.

What would you do if you were giving an important presentation and the projector bulb suddenly burned out? I always have a hard copy of summary notes, and I welcome the challenge of trying to paint the picture verbally.

As the drummer in a band, what would you do if your snare drum head (the skin) broke suddenly? For important gigs, I have a spare head ready to be rigged. When I don't have a spare, I am prepared to flip the drum and play the snare side. By the way, in forty years of playing I have never broken a head. But I always expect that day is coming.

Assume a failure will occur at the worst time. Develop redundancy and contingency plans. Adopt the operator's negative polarity thinking to your benefit. Embrace constant wariness, healthy paranoia, and eternal vigilance. Never forget how dangerous your job is.

When testing equipment, always assume the test will fail. Know what to do immediately if the test does fail. Astronauts harbor this negative mental outlook when conducting the vestibule pressure checks before opening the hatch between the docked Space Shuttle and Space Station.

The same mental attitude is valuable in the oil-and-gas industry. Good drillers assume the negative-pressure test will fail when testing the cement barrier prior to removing the heavy drilling fluid from the drill pipe before disconnecting from the well. This assumption of failure is advisable especially because the test itself is designed to encourage the well to flow. The driller should be anticipating the need to take immediate corrective actions to control the well during the test.

When driving, I always assume other drivers are trying to kill me. When I push the ignition button in my car before driving, I try to visualize my broken, bloody body after an accident I failed to prevent. So far, this technique has scared me into successfully preventing accidents (with help from a few other defensive techniques).

As a pilot, I always expected the ground controllers would make mistakes, and I was ready to help them rectify their errors. Later in my career, when I flew in an airplane with another crewmember, I usually mentioned in the preflight briefing that I would make a mistake sometime during the flight, unintentionally, of course. My crewmember's assignment was to identify the error quickly.

When flying a single-engine airplane, I had an automatic subroutine in my brain that constantly calculated what I would do as soon as the engine failed. It never happened, but I was always ready.

Two things are certain about failures. First, if you expect failures, you won't be surprised if they occur. Second, whether you expect failures or not, they will occur.

Example: Arrested Landing Aboard an Aircraft Carrier

Naval aviators learn early in their careers to expect failures. The technique of expecting failures is used during every arrested landing aboard aircraft carriers (shown in photo 75). Sometimes, the tailhook fails to engage one of four arresting cables. At less

Photo 75: Arrested landing aboard USS *John F Kennedy*; the pilot selects full power on the engine, anticipating the hook/cable system will fail to arrest the landing.

frequency, the tailhook improperly engages a cable, which begins to decelerate the airplane, but before the velocity reaches zero, the cable is "spit out." Rarer still, the cable snaps because of a mechanical defect and overload.

The causes don't really matter during the arrested landing attempt. Every naval aviator on every arrested landing selects full power on the engine(s) immediately after main wheel touchdown. The engine throttles are never retarded until a flight-deck crewmember gives a hand signal. Every failure scenario is anticipated, and if any occur before the airplane is safely stopped, the engines are already screaming at full power, trying to help regain sufficient flying speed. If they can't, the airplane will be jettisoned into the ocean under the pilot's ejection seat.

Example: Catapult Launch
Now, think about the launch sequence at the beginning of flight from an aircraft carrier. What failures do we expect? How does the technique of expecting failures change our thought processes during a catapult launch, as shown in photo 76?

Remember the engine has just been started from a cold state, as recently as ten minutes ago. Seconds before the catapult is fired, the pilot applies full military-rated-thrust (or maximum afterburner) power to the engine(s), which spools up to 100 percent rpm with maximum fuel flow and exhaust gas temperature. In addition to the sudden rotational and thermal stresses from high rpm and temperature, longitudinal stress to the engine components is off-the-charts high due to the translational acceleration from 0 to 150 knots in two seconds. Remember the vulnerability of hardware in transition? If the engine is going to fail during the flight, now is the likely time, especially on a stormy night, just as the launch catapult is shocking

Photo 76: Catapult shot from the USS *John F Kennedy*

the neurovestibular system of the pilot and causing some amount of (expected) disorientation.

On every launch from an aircraft carrier, I always anticipated some failure would cause my airspeed to be insufficient by the time I reached the end of the flight deck. During every one of my 345 catapult launches, all the way down the stroke, I fully expected to eject. The reason did not so much matter. It could have been because of an engine failure, a cold shot from the catapult, a launch bar failure, or any other failure resulting in insufficient airspeed. I was ready to eject from the airplane and go swimming, day or night. By the way, I also expected an airspeed indicator failure, so I had to execute a verification strategy to assess my actual airspeed before deciding to eject.

In every case, I experienced no failures during launch, and, with sufficient flying speed, I executed my *secondary* plan. I stayed in the airplane and kept flying.

***

Example: A-6 Arrival at Roosevelt Roads; T-38 Departure from Moffett Field
Here is a two-part story I use to illustrate the dangers that can happen when pilots do not expect failures.

In the late 1970s, I was sitting in the pilots' ready room at Naval Air Station, Roosevelt Roads, Puerto Rico. An irate and seemingly irrational pilot stormed into the ready room with his bombardier-navigator after landing in their A-6 Intruder from a sister squadron in Carrier Air Wing One. The incensed pilot threw his kneeboard across the room into the far wall and screamed, "That damned controller tried to kill me!"

The reason for his emotional explosion was this. With the weather deteriorating, the pilot was receiving vectors (verbal commands for airplane heading) on the radio from the approach controller, who was using ground-based radar for positioning the airplanes on final approach to landing. A similar approach in low visibility conditions is shown in photo 77, taken from my single-seat A-7 on the same deployment at Rosie Roads.

Photo 77: On final approach, receiving directional guidance from ground-based approach controller, NAS Roosevelt Roads, Puerto Rico

The A-6 pilot was given a vector that had him headed straight toward a mountain, which was obscured by the clouds, prior to being turned onto the final approach course for the runway. This is the normal and expected approach to this airport. Unexpected though, the approach controller became temporarily distracted and left the A-6 on the heading toward the mountain, overshooting the extended centerline of the runway. After recognizing his error, the approach controller avoided disaster by giving the overly trusting crew a corrected heading back toward the centerline of the runway.

Two thoughts went through my rookie new guy head after hearing the enraged pilot's complaint. First, the A-6 is an all-weather capable airplane with twice as many crewmembers as I have, who are supposed to be able to fly blind through mountainous terrain at low altitude. Second, ground controllers can "try to kill me" as much as they want; they will be unsuccessful because avoiding mountains I know are hidden in the clouds is my responsibility, not theirs.

Always expect failures—in systems and humans. Don't let either kill you.

The second part of this story occurred two decades later in the pilots' ready room at McCarran Field in Las Vegas, Nevada. By the mid-1990s, I was an astronaut at NASA flying the T-38 aircraft. We had recently upgraded the avionics in the T-38 and retrofitted the cockpit with advanced displays. This "glass cockpit" improvement made the T-38 a safer airplane, more automated, and more capable of navigating in bad weather with no visibility. A picture of a newer version of the cockpit is shown in photo 78.

Photo 78: T-38 glass cockpit with Flight Management System navigation based on GPS

The centerpiece of the avionics upgrade was a Flight Management System (FMS), which received information from Global Positioning System (GPS) satellites and automated the navigation task for the pilots.

As I was sitting in the ready room, another NASA pilot stormed in, slammed his kneeboard on the desk, and screamed, "That damned system tried to kill me!"

I learned he departed from NASA Moffett Field, in San Jose, California, en route to McCarran Field. As he flew along the departure route to avoid the mountainous terrain in bad weather, he was following the automated FMS guidance information in his T-38. The FMS momentarily lost track of the GPS satellite and left the pilot on a heading directly toward a mountain obscured in clouds.

Again, two things went through my mind as the pilot screamed. First, *Why are you letting the system try to kill you?* You should understand the failure modes in the automation, expect failures to occur, and know what to do to operate safely with any failure. Second, *Some things never change.* Twenty years ago, some pilots allowed approach controllers to "try to kill" them. Now, some pilots are allowing automated systems to "try to kill" them.

In this technique, *Expect Failures (System and Human)*, I will offer some thoughts on the relative strengths and weaknesses of computers versus humans and the value added and challenges created by automation of systems.

I am a fan of automation. Computers have helped operators by decreasing their workload. With advanced technology, automation has helped humans perform better and more accurately to achieve greater mission results. Dangerous operations are made safer with the precision of computers and the perseverance of automation.

With all the advantages, there are some disadvantages to automation. Humans and computers both fail. Most of the time computers perform better than humans. Sometimes the computers crash, taking the humans with them. Since pilots don't like crashing, we are reluctant to allow computers to control critical operations without human oversight and override capability.

So how do we balance the upsides and the downsides? In space, astronauts allow computers to do what computers do best. Because computers never get bored or tired, we allow the computers to conduct continuous, tedious, or mundane tasks. Computers handle tasks that require accuracy or efficiency, but only if the tasks can be programmed accurately and completely. Attitude control is almost always relegated to the autopilot. Computers monitor the status of critical systems and annunciate failures when detected by sensors.

We do what we do best. Humans use judgment, experience, and intuition to make decisions and take actions when information or training is incomplete. Humans control systems when the control task can't be programmed accurately and completely into a computer, or when failures cannot be tolerated. This is why American astronauts controlled the Space Shuttle manually when docking to the International Space Station and performed the landings manually on the runway. We couldn't afford to have a computer glitch when docking or landing a one-hundred-ton glider. (The Russian Space Agency uses a different philosophy. Russian cosmonauts usually allow their autopilot to dock the Soyuz vehicle to the Space Station automatically. Their vehicle is smaller and more nimble than the Space Shuttle was, and they can safely take over manually if their automatic system fails.)

Humans oversee the computers for one important reason. Computers don't have a fear of death. If a computer makes a mistake, it can lead us to destruction, without missing a byte. Actually, computers do make more mistakes in space than they do on Earth, though the fault is not necessarily theirs. Without the radiation protection normally afforded by the atmosphere, the computer's memory is vulnerable to stray particles from the sun that cause single-event upsets, which may flip a bit, resulting in flying *into* the moon instead of *to* the moon (figuratively, but you get the idea). The human's neural network is designed differently and is apparently much more robust, at least for several decades. I've noticed some degradation in the robustness and accuracy of my brain now that I'm more than sixty years old. But I still have a healthy fear of death. I rely on my old brain to keep my young computer and me alive.

So we use computers where computers are best and humans where humans are best. With the rise of computers, automation provided enhanced safety and productivity by increasing precision and decreasing workload. But read the devil's fine print closely. In a dangerous operation, automation can kill quickly. Automation software is complex, so bizarre and obscure failure modes are possible—and harder to identify. The necessary corrective actions are more challenging to execute, especially in hazardous conditions where small errors can cascade quickly to catastrophe. Though the workload for the operator may decrease during operations, the preparation workload for the operator *before* the operation increases significantly. Good operators understand automation, how it changes risk, and their added responsibility to monitor the computer's performance.

Always expect failures. With automation, expect insidious failures. To control risk and stay alive, the best operators understand complex, subtle, and strange failure modes, and how automation will function during degraded modes of operation. Most importantly, the best operators understand how and when to take over from the computer to save the crew, the vehicle, and the mission. That's what we humans do best.

Example: Close Approach to Russian Space Station Mir *on STS-63*
During preflight training, the crewmembers of STS-63 learned the mission objectives had changed. Russian and American program managers realized STS-63 could be a test flight in preparation for the first docking between an American Space Shuttle and the Russian Space Station *Mir*, scheduled to occur three months later. We were tasked with conducting an operational test and evaluation of the flight procedures and systems for rendezvous. Additionally, the test was an important checkout of the ground procedures and systems used in the Mission Control Centers coordinating with each other in two languages across the Atlantic Ocean. Our job was to fly *Discovery* to a position ten meters away from *Mir*. A picture taken during the close approach is shown in photo 79.

Photo 79: Dr. Valeri Vladimirovich Polyakov on Russian Space Station *Mir*, in a photograph captured by Dr. Bernard Harris from the Space Shuttle *Discovery* during close approach, STS-63 (Photo Credit: NASA)

At one point during preflight training, our crew, which included Russian Cosmonaut Vladimir Titov, flying with us as a mission specialist, was invited to a meal with the most senior managers in the Russian space program. During our conversations, I learned the venerable and experienced Russian managers were quite worried about our flight. No doubt, they were engaged in the Russian version of the technique Expect Failures *(in the human—me!)*. They kept looking at me, a young pilot with unknown flying abilities, wondering why they were allowing me to fly a massive, 120-foot-long spaceship right next to the pride and glory of the Russian space program, the beautiful *Mir* Space Station.

After landing, I learned why they were so concerned. The *Soyuz* capsule has a much higher thrust-to-weight ratio than the Space Shuttle. If the *Soyuz* is the racecar of space vehicles, the Shuttle was the eighteen-wheeler, or the ocean liner. The Russian managers only had the experience of seeing the much-less-massive *Soyuz* close to *Mir*, as it danced and jittered nervously under the control of the high-thrust autopilot system. Unknown to them before flight was the Space Shuttle, designed by brilliant engineers, could be flown smoothly and precisely to within an inch of the desired position and within hundredths of a foot per second of relative velocity. The test flight of STS-63 gave program managers, on opposite sides of the planet, confidence in the procedures and the predicted success of the first docking flight three months later.

Accuracy of the position of *Discovery* relative to *Mir* was critical during the close approach. To verify our position, we used four dissimilar systems to measure or calculate relative position—Rendezvous radar, hand-held Laser Illuminated Detection and Ranging system, navigational computation using General Purpose Computers, and hard-mounted optical reticle ("iron gun-sight").

Expect failures. Assume they will occur at the worst times. Just prior to the close approach, we had a two-on-two split of sensors. Two of the sensors agreed and calculated an identical relative position. The other two agreed also, but had calculated a different relative position than the first two sensors.

Our immediate operational task was to resolve the dilemma safely. As a pilot, I wanted to make an accurate decision. But, here's the thing about making operational decisions in a dangerous environment. As a *test pilot* on this mission, my overriding responsibility was to make the decision that was the more conservative and safe, even if inaccurate and resulting in not accomplishing the objective. Our overriding flight rule specified we were to fly no closer than ten meters between any parts of the structure of either vehicle. This meant we had to choose the sensors that showed us to be at the closer of the two choices. If postflight analysis showed the chosen sensors were actually in error, then, at closest approach, *Discovery* would have been farther away than ten meters from *Mir*. So, at least we would have conducted our operation safely.

To close out this story, this is what occurred next. As planned, we continued the approach using the two sensors that gave us the conservative solution. Coincidentally, one of these two sensors was the simplest and most basic system, the iron gunsight (shown in photo 80, on another flight). The other three sensors were more complicated systems with many possible failure modes. Not much can go wrong with a simple, optical reticle hard mounted to the structure of the vehicle.

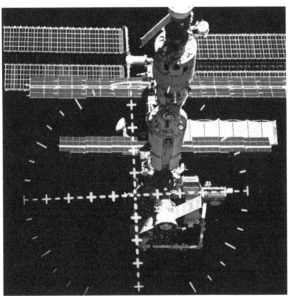

Photo 80: View from the Space Shuttle during close approach to the International Space Station on STS-105. A similar hard-mounted, optical reticle was used to resolve the disagreement of two-vs.-two sensor split when flying to *Mir* on STS-63 (Photo Credit: NASA).

As we continued our conservative approach, Mike Foale, our navigator and one of the smartest people I know, sleuthed out the problem. Previously, we erroneously entered a single parameter into the navigation system that corrupted the calculated position from two of the four sensors. In this case, the expected failure was our own human error. He corrected the input, and by the time we arrived at closest approach, all four sensors agreed. We conducted proximity operations with *Mir* for ten minutes, not one centimeter closer than ten meters.

There is one final caveat to this story, appropriate for this technique of expecting failures. We anticipated one other failure by the commander (me) might happen at the worst time. To prevent damage to *Mir*, the attitude control jets on *Discovery* were selected to fire only in directions away from *Mir* when we were within 1,000 feet. This meant, if unchecked, *Discovery* always had a tendency to accelerate toward *Mir* during proximity operations. Mike's assigned responsibility was to monitor distance, in the event I became distracted when giving the requisite feel-good speech, once in Russian and once in the English. If you listen to the downlinked audio of my speech, you may hear Mike in the background commanding me to, "Pull out, pull out!" As anticipated, I was distracted, but Mike's commands prevented a violation of the flight rule. Expecting failures and anticipating corrective actions saved our day. Twice.

9. Develop Error Wisdom (Individual and Collective)

Humans learn quickly on the edge of failure. The mistakes people make can foster the best learning to acquire a skill. We really do learn from mistakes. But in hazardous environments, mistakes can kill so there's the challenge. Mistakes give us great opportunities to learn, but we need keep those mistakes small enough to stay alive to learn again the next day.

Avoiding errors and their resultant bad consequences is the purpose of the technique after this one. But, before we can develop effective techniques to avoid errors, we need to understand how errors are made. Interestingly, one of the most important lessons we can learn from this understanding is appreciating how easy making errors is.

Each individual has a personal propensity for making errors. To be successful in hazardous environments, operators should understand how their brains work, determine their individual error-making propensity, and develop personalized techniques for preventing errors and their consequences. Psychologist and author James Reason uses the term *error wisdom*.[20] Operators should develop error wisdom to understand how errors are made and how to develop and use techniques to avoid adverse consequences from errors made so easily.

The following example illustrates how error wisdom could have been beneficial. Of course, wisdom is only useful if used wisely.

Example: STS-63 Crew Patch (Approved)

By 1994, I had been flying as a naval aviator for eighteen years. I had developed sufficient error wisdom to keep me alive and enable me to improve my operating performance in hazardous environments. But, I had not yet appreciated the power of developing error wisdom in nonhazardous operations or in administrative tasks. That changed before STS-63.

---

20   James Reason, *Beyond the Organizational Accident: the Need for "Error Wisdom" on the Frontline*, December 2004 (Quality Safety Health Care); http://qualitysafety.bmj.com/content/13/suppl_2/ii28.abstract.

Before every spaceflight, each crew is responsible for designing the crew patch for their mission. Some crews solicited the help of artists or designers in developing their crew patch. The commander of the mission selects a design and submits the desired version to NASA headquarters for final approval.

On the six missions I flew, one or more members of our crew designed each patch. Before STS-63, our crew went through several designs and modifications before converging on one final design. I asked the graphic designers to review the final version one last time before I submitted my recommendation for approval. Photo 81 shows the version of our crew patch I recommended to NASA headquarters for approval. I should have used a process with sufficient error wisdom to identify and eliminate any mistakes that might have been made. I didn't. Can you identify the error in the design of the flags in this version of the STS-63 crew patch I recommended for approval? The answer is in the footnote at the bottom of this page.[21]

Photo 81: Version of the STS-63 crew patch the commander submitted for approval. Can you spot the error?

---

21 Stop reading now if you want to continue searching photo 81 for the error in the STS-63 patch. We inadvertently included two extra rows of stars in the American flag. Usually, we insert hidden meaning in various aspects of our crew patches. Giving us undeserved credit for always operating error-free, some gracious supporters of our space program wrote to us and asked, "What's the significance of sixty-one stars?"

We learn from making mistakes. Some people learn better and more quickly than others. Some have more opportunity to learn than others. Given the number of mistakes I have made, if I truly learn from my mistakes, someday I'll be brilliant. This process of learning from mistakes and becoming skilled occurs automatically through practice. But in dangerous endeavors, operators don't have the luxury of waiting for the automatic process. We must accelerate and deepen the learning. To perform at a high level of excellence, operators must deliberately and skillfully learn from mistakes. The best operators analyze their mistakes. From these analyses, the best operators devise personal plans and techniques to reduce errors, capture incipient errors, and minimize the consequence of the errors they make.

Example: Press Conference from Orbit
Here is another example of an error we made while conducting an administrative task. Near the end of the joint phase of mission STS-102, we conducted a press conference with the media in the US and Russia, which was televised to the various participating nations involved in the partnership constructing the International Space Station (photo 82).

Photo 82: Joint crews of Expedition-1, Expedition-2 and STS-102 in the *Destiny* laboratory module of the International Space Station on the day of our press conference. (Photo Credit: NASA)

In the hours before the press conference, crewmembers prepared the *Destiny* module for the live video downlink, including configuring the video and communication equipment and installing the flags of the participating nations on the bulkhead behind the crew. Though this was an administrative task involving almost no hazards, the astronauts or cosmonauts had to conduct this simple, but detailed, task in a hazardous environment, with some low-level amount of associated physical and psychological stress.

Error wisdom tells us humans could easily make mistakes while conducting tasks with many small details under time pressure or psychological stress. Error wisdom would have helped us design a process to verify all errors in displaying the national flags were identified and corrected before the start of the press conference that included live video transmission to the ground.

I don't know if the crewmember who installed the flags succumbed to adverse time pressure or psychological stress in the hazardous environment, which resulted in unintentionally displaying one flag incorrectly. I do know none of us possessed or used sufficient error wisdom to foresee the possibility of this error and to design a process to help us avoid the mistake. Can you identify which flag was upside down in photo 82? Unfortunately, before the televised press conference, we didn't. The answer is found in this footnote.[22]

This technique, *Develop Error Wisdom,* helps us prepare to learn from errors. Error wisdom can be developed from the following learning opportunities:

- Acquiring valuable experience in preventing errors or keeping errors small with limited consequences, through progressively increasing responsibilities in hazardous operations
- Sage advice from instructors, coaches, mentors, and operators with more experience
- Self-reflective, highly introspective, honest, self-critical analyses of error patterns and evaluations of operating performance

The last set is critical. You are the only one who knows the magnitude and the details of the errors you have committed or were about to commit. Instructors can help, but your own honesty is the key to developing deep, personal error wisdom. Analyses of the error patterns of an individual or group can indicate creative solutions to help managers and operators reduce the amount and magnitude of errors committed.

Example: Self-analysis of Commander's Errors During Preflight Training for STS-113
In 2002, I began training for my final mission in space. With the privilege of having flown five missions, I had little excuse for a sixth less-than-perfect mission. Two other factors added to my self-imposed pressure to strive for a flawless flight.

---

22   Regrettably, we erroneously displayed the German flag upside-down during our joint press conference from the *Destiny* module.

As the director, Flight Crew Operations, I was responsible for codifying the Principles of Operations for spaceflight crews and for leading our effort to develop the Techniques for Operating Excellence. This flight would be a demonstration of the effectiveness of all the techniques I learned in twenty-seven years as a naval aviator.

Finally, as this would be my last mission, I knew I would have to live with the consequences of errors I made on this flight for the rest of my life, whether my life ended peacefully fifty years later or violently during the flight because of a mistake I made.

For these three reasons—to justify the privilege, to prove the techniques, and to stay alive—I intended to drive our error rate as close to zero as possible, while making no errors that had any operational consequence. I started with my own performance. For most of my flying career, I had been making more mistakes than I cared to admit. With increasing experience, I was able to reduce the number of errors, trap impending errors, and minimize the consequences of errors I made. But the number was still greater than I could afford on my final mission.

I began to analyze my errors in training, with the intention of creating a strategy to eliminate, or at least mitigate, my errors. My first task was to record the errors I made in the Space Shuttle simulator. A subset of my list of errors in training is shown as a partial spreadsheet in Table 5. I decided to include mental errors I almost made but didn't, thanks to another crewmember or flight controller who said something to preclude my impending error.

DATE	SYSTEM	MISTAKE	CRIT
20 Aug 02	CONTROL	High NRG Gander: After bug < Thermal line, didn't reduce pitch fast enough	1
12 Oct 02	CONTROL	Low NRG Cherry Pt: Low to High (didn't follow Guidance after App/Lnd)	1
12 Aug 02	DPS	BFS recovery: Didn't complete AESP [pro to OPS 3] (TWICE)	1
31 Oct 02	DPS	BFS recovery: Didn't recognize recovery [pro to OPS 3] (TWICE)	1
13 May 02	DPS	BFS: 102 to 601 (w/o 103)	1
20 Aug 02	DPS	FA2 dn: Didn't take FCS Ch to OFF	1
27 Jun 02	DPS	IMU: Didn't notice GPC3, IMU2, FF1 fail; Didn't restring, Reselect IMU3	1
4 Jun 02	DPS	Set Split took 15 sec to see; (SSME dn simo)	1
12 Aug 02	DPS	Set Splits: Took too long to get BFS Sys Summ display	1

6 Aug 02	DPS	Strg 1 2 3 3: Didn't recog I/O Term B, [Look at 3 dn arrow]	1
12 Oct 02	FDF	3 Out blank: Didn't select TAL	1
12 Aug 02	GUID	Cue SERC: Misdiagnosed R SSME dn, APU1 dn [Exec if APU 2 or 3 dn]	1
12 Aug 02	GUID	Delayed TAL: Invoked at new no-comm bndry before MCC call	1
4 Jun 02	OMS	Multi DPS ASC: didn't do Man s/d L OMS	1
26 Jul 02	RNDZ	Low Z: F1 Manif dn, GPC 3 dn, Selected AUTO before restring	1
22 May 02	BFS	BFS PL1 dn: Missed I/O Term A dn [Use Ack key if CRT in use]	2
4 Jun 02	BFS	Post Bus Tie: Didn't do I/O Reset to BFS	2
27 Aug 02	DPS	BFS I/O Term B failed low: Didn't pause in Norm (tb); didn't realize why Term	2
13 May 02	DPS	BFS: stuck in 102 (strg 3 & 4 dn), didn't see	2
20 Aug 02	DPS	FA2 dn: After Port mode recovered, didn't do I/O Rst to BFS	2
6 Aug 02	DPS	FF1 Pwr, No comm: Incorrectly did pwr cycle (w/o "go")	2
12 Aug 02	DPS	I/O Term B: Restring, asked MCC if Stby/Halt needed	2
27 Jun 02	DPS	Set Split, Set Split: Misdiagnosed signature	2
6 Aug 02	DPS	Strg 4 trade case: Stayed on FF4 (Ent)	2

Table 5. Partial spreadsheet showing an analysis of training errors the STS-113 commander committed over six months in the simulator before the mission

Creating the list resulted in my first lesson from this process. I realized I was making errors in my thought processes that neither my crewmates nor the instructors knew about, since someone else fortunately prevented my impending error with a timely statement or action. If I wanted a complete list of my errors, I was the only one who could create the list. The professional instructors or flight controllers couldn't help, as good as they were, because they could not see the occasional erroneous thinking going on inside my helmet.

For the next six months, I recorded every error I made in the simulator that resulted in a consequence of losing the crew, vehicle, primary mission, or a secondary objective. This list resulted in the second lesson I learned from my process of analyzing personal errors. Over the six-month period, I made more mistakes than I had predicted. In any given four-hour session in the simulator, the instructors could drive us to our knees with overwhelming intellectual and emotional stress from the devious, complex operational scenarios with system failures piled on top of deadly flight environments. I loved every minute of it. But, with no fewer than two four-hour sessions per week, every week for six months, I had a large enough sample size of errors to conduct a valuable analysis of my error-making trends.

In analyzing my pattern of errors, I was interested in answers to two questions. First, was I learning from the training and making fewer errors as the syllabus progressed? Second, how many errors was I making with high consequences?

Using the spreadsheet, as shown partially in Table 5, I logged each error in a row, with columns indicating the date, the system in which the error was made, the details of the error, and finally, the criticality of the error. I created definitions for the levels of criticality. Crit 1 meant my error caused the loss of the crew, the vehicle, and the mission. Crit 2 was reserved for errors causing the loss of the vehicle and the mission, but we saved the lives of the crew. Crit 3 meant we failed to complete the primary mission but saved the crew and the vehicle. The least critical, Crit 4, indicated we completed the primary mission objectives but failed to complete one or more of the secondary objectives.

To answer my first question, I simply sorted the rows by date. From this analysis, I could see as the training progressed, I was making fewer mistakes over time. As always, the training improved my confidence as we approached the launch date.

In analyzing the data to answer my second question, I learned the more important—and unexpected—lesson from this process intended to develop error wisdom. I was interested in understanding how many errors I was making that had high consequences. What I learned, though, was even more valuable. My critical errors were in a different system than I was expecting. When I sorted the rows by level of criticality, I could see my performance was much worse than expected as I made simple keyboard entries into the flight computers (labeled DPS in the chart).

This was a system in which I previously had high confidence in my capability. My weaker performance when using this system didn't register in my mind before the analysis because errors in keyboard inputs didn't usually result in immediate loss of vehicle and crew. The death and destruction usually occurred much later during the reentry as the orbiter was trying valiantly to save itself, but couldn't quite solve the impossible aerodynamic situation I had created twenty minutes earlier with my erroneous inputs into the computers.

Had I not conducted this analysis of my errors over a six-month period to identify trends, I never would have realized I needed to improve the way I was operating with a particular system. This analysis gave me a deeper understanding of my errors. With this newly developed error wisdom, I was able to create a training strategy for my crew and me to improve our operating performance.

———

What operators do with the individual and collective error wisdom they develop is the purpose of the next technique.

10. Use Error-Mitigation Techniques

In the 1980s, the aviation industry realized something needed to be done to help pilots reduce the number of accidents caused by human error. A NASA psychologist, John Lauber, used the term *Cockpit Resource Management* to describe the process pilots adopted to improve performance in the cockpit. The name of the process was later changed to Crew Resource Management.

One of the powerful elements in the process of Crew Resource Management is the concept of the Error Troika,[23] developed by Robert L. Helmreich from the University of Texas at Austin. The Error Troika, listed as A, B, and C below, is the basis of this technique, *Use Error-Mitigation Techniques*.

After analyzing errors and developing error wisdom, the best operators identify personal techniques to eliminate errors committed during various operations or, at least, reduce the likelihood of errors being committed. Knowing all humans make errors, the elite operators strive to develop and use techniques to capture incipient errors and mitigate the consequence of any errors that have occurred in all phases of operations.

To help me remember these three error-mitigation techniques, I use a memory aid described in some of the previous techniques by thinking of the timeframes of past, present, and future. Operators should develop and use error-mitigation techniques to:

A. Reduce the Likelihood of Errors (*Before* the Operation)
B. Capture Incipient Errors Before They Occur (*During* the Operation)
C. Mitigate the Consequences of Errors (*After* the Operation)

A. Reduce the Likelihood of Errors (Before the Operation)

The primary and most effective subtechnique used to mitigate errors is to *Reduce the Likelihood of Errors* being committed in the first place. The organization and individual operators can use various methods to decrease the probability of errors

---

23 Robert L. Helmreich, Ashleigh C. Merritt, and John A. Wilhelm, "The Evolution of Crew Resource Management Training in Commercial Aviation," International Journal of Aviation Psychology 9:1 (The University of Texas at Austin, 1999), 19–32; also http://lessonslearned.faa. gov/L1011Everglades/crmhistory.pdf.

being committed. Training is the most obvious method. When humans become skilled, they are less likely to commit errors. The use of checklists, procedures, and a lock-out/tag-out process are other ways to reduce the likelihood of errors in operations.

A more effective way to reduce the likelihood of errors is to design the system using Human Factors standards. Human Factors is the study of how operators interface with systems. The intent of the standards is to help the designer create systems that optimize the performance of the humans who are using the system, making it less likely operators will commit an error.

In the technique Control Risk, I mentioned an example of a design improvement that decreases errors made with the breathing systems in hospitals. Originally, the oxygen and nitrogen hoses had common fittings to attach to a wall-mounted interface in the breathing devices for the patient. Fatal errors have been committed when a nitrogen hose was connected to an oxygen port. So the system was redesigned to incorporate different and incompatible fittings. As described in *Doctors' Errors and Mistakes of Medicine: Must Health Care Deteriorate?*, "Oxygen attachments are screwed turning in the opposite direction to those of other gases."[24] Nurses are no longer able to connect a nitrogen hose erroneously to an oxygen port.

Unfortunately, as I presented in chapter 2, individual operators don't have the luxury of designing or redesigning the hardware or software in the system to optimize human/system performance. Operators make more mistakes when using poorly designed systems. Here are some quick examples to illustrate designs that do not contribute to optimized performance.

Example: Upside-down Numbering of Floors in a Hotel
In 2007, I stayed in a hotel built into the side of a hill in the city of Port-of-Spain, Trinidad. Oddly, the numbering system used to designate the floors in the hotel was upside down. This confusing system is shown in photo 83. For the entire two weeks during my stay, I made multiple errors trying to navigate in the hotel. I never was able to devise a mental method to eliminate my errors. I imagine this unconventional system might introduce additional hazards when a quick evacuation is required during an emergency, because the only exits from the building to safety were on floors 5 and 12.

---

24   Moshe Wolman and Ruth Manor, Doctors' Errors and Mistakes of Medicine: Must Health Care Deteriorate? (Burke, VT: IOS Press, Inc., 2004), 53.

Photo 83: Elevator configuration with the upside-down numbering system for the floors in a hotel built into a hill in Port-of-Spain, Trinidad

Example: One-way Door with Pull Handles on Both Sides
Photo 84 shows a double door in the lobby of an office building. This set of doors can be opened only in one direction. As you can see through the glass, the doors have been designed with pull handles on both sides. From the side where a push is required, many people pull on the handle, attempting (unsuccessfully) to open the door. A better design would have a push plate, rather than a handle, installed on the side where only a push will be successful.

Photo 84: Pull handles on both sides of a door in the lobby of an office building

With insight into how humans generally interface with systems and with error wisdom developed from self-analysis of how errors are made, an operator can devise methods to reduce the likelihood of committing an error. Creating such methods is the basis of this subtechnique. Here is an example of creating a method to reduce the likelihood of errors.

Example: Reducing the Likelihood of Medical Errors
The medical community has a long history of operating teams making errors during surgery. In the extreme, patients have woken to find the wrong limb amputated.

To study and understand the quality of health care, the Institute of Medicine, established by the National Academy of Sciences, formed a Committee on Quality of Health Care in America. In 1999, the committee issued a report, *To Err Is Human: Building a Safer Health System.*[25] Unprecedented in the health-care field, this report

25   Linda T. Kohn, Janet M. Corrigan, and Molla S. Donaldson, editors, *To Err Is Human: Building a Safer Health System*, Committee on Quality of Health Care in America, Institute of Medicine

was influential in helping doctors develop error wisdom and begin to change the culture in operating rooms.

Surgeons and their operating room staff introduced simple techniques to reduce the likelihood of operating on the wrong body part with bilateral symmetry on the patient, but some initial attempts were not completely effective. Teams considered using a technique of drawing a big *X* with a sterile marker on the correct body part intended to indicate, "X marks the spot,"[26] operate here. Unfortunately, this technique could lead to ambiguity, as some doctors might misinterpret (understandably) the big X to mean, "Wrong side; do not operate here."

The fix was simple. Now, some surgical teams write the word "Yes" on the correct body part. Even better, some surgeons write their initials at the correct location. To further limit the likelihood of making a mistake, the word or initials are printed with the concurrence of the patient before the anesthesia is administered. Photo 85 is a picture of my wife in the hospital during pre-op on the day of her shoulder surgery in 2010. Note the word "Yes," printed with a marker on her left (correct) shoulder. For this surgery, her left shoulder was the right shoulder, and her right shoulder was not the right shoulder. We are glad the operating staff was not confused.

The other symbols written on her body were notes to the operating team. As long as the notes were unambiguous and relevant, we were happy for them to write anything they wanted to help the surgical team avoid errors.

Photo 85: Clinical technique used before a medical operation to reduce the likelihood of an error being committed during a procedure on my wife, Robin.

(Washington, DC: National Academy Press, 2000).

26   Joint Commission on Accreditation of Healthcare Organizations, *Issues in Provision of Care, Treatment, and Services for Hospitals* (Oakbrook Terrace, IL: Joint Commission Resources, 2004), 72.

B. Capture Incipient Errors Before They Occur (During the Operation)

Humans are fallible. All operators will make mistakes. In dangerous operations, some errors cannot be tolerated. To prevent accidents, these critical errors must be trapped. Again, with the proper error wisdom, developed from the knowledge of how people make mistakes, operators can devise methods to *Capture Incipient Errors before They Occur* during operations.

The two-person rule is an effective technique that can be used to capture incipient errors immediately prior to the operation. When used correctly, one operator will communicate the intent of an action to a second operator. This gives the second operator an opportunity to prevent a potential error about to me made by the acting operator.

A single operator may exercise this subtechnique in other ways. A verification step can be inserted into a procedure to instruct an individual operator to identify and prevent potential errors immediately prior to execution. Various error-correction schemes can be used when working with spreadsheets or software programs to prevent errors during activation or operation.

Example: Rendezvous Operations in Space (and Parenting Activities on Earth)

Rendezvous operations in space are complicated, though not complex. Atul Gawande draws a nice distinction between complicated and complex problems in his book, *The Checklist Manifesto*.[27] A rendezvous between two orbiting bodies in space involves many teams and thousands of interrelated, interdependent activities that must be conducted in precise, integrated sequences to create success. The operation is complicated, but the activities are repeatable, and success can be achieved predictably if the operators don't make mistakes with adverse consequences.

Complex problems, such as parenting, are even more difficult. Tasks and activities are not necessarily repeatable. Outcomes are unknown, even when following a process, and expertise may not be sufficient to generate success under such uncertainty.

In either complicated or complex operations, humans are prone to make errors. There are so many interdependent tasks that performing without error seems quite challenging. When complicated operations are also dangerous, one small error can result in cascading mistakes and quickly escalating danger. In the hazardous environment of space, with vehicles orbiting at extreme velocity and high mass, one mistake can result in disaster.

So, in either complicated or complex problems, operators need techniques to Capture Incipient Errors Before They Occur. As you might expect, one of the best

---

27 Atul Gawande, *The Checklist Manifesto* (New York, NY: Metropolitan Books, Henry Holt and Company, LLC, 2009), 49.

techniques to use in complex parenting activities, the two-person rule, also works in complicated rendezvous operations. But in rendezvous operations, tasks are so critical and dangerous that operators cannot afford to make even a small mistake that goes unchecked. So the two-person rule is expanded to become the two-team rule. Photo 86 shows us using the two-team rule to prevent errors during dangerous operations in space.

Photo 86: The two-team rule (more effective than the two-person rule) is used during complicated rendezvous operations in space to capture incipient errors before they occur. (Photo Credit: NASA)

Before all critical maneuvers during our rendezvous operations, a common question we asked ourselves was, "What mistakes have we made earlier that might cause this maneuver to fail?" Simultaneously, our ground control team was asking the same question. On my final mission, we found this was the most powerful method to capture errors before they occurred. Often, the ground and flight teams found no previous latent errors had been introduced before a critical activity. Sometimes, though, we did find errors that needed to be corrected before the activity was executed. But no error was able to escape the scrutiny of this final question to survive long enough to affect the operation by causing any adverse consequence.

C. Mitigate the Consequences of Errors (After the Operation)

During operations, sometimes errors are committed, even after using methods to eliminate or reduce their likelihood. After an error has already been committed in a dangerous operation, the only option remaining is to identify the error quickly and take corrective actions to reduce or eliminate the consequences of the error.

This is where pilots excel. Good pilots use the best methods to minimize the likelihood of committing errors and capture them before they are committed.

But, on the rare occasion they do commit errors, good pilots are exceedingly fast at identifying the error and correcting it effectively to minimize any adverse consequences.

The best operators identify all errors before any adverse consequences can be manifested. Elite operators are willing to admit quickly to teammates when they have made errors. Their admission allows the team to help minimize the consequences.

Example: Backup Hard Drive
Here is an example of a simple and effective device that can be used after an operation to *Mitigate the Consequences of Errors* made during the operation. Two kinds of people live in our modern world—those who have never been frustrated with computers, and those who use computers. Routinely saving important files on a backup hard drive, as shown in photo 87, can help the user mitigate the consequence of errors made at the worst possible time, just before the file disappeared.

Photo 87: A backup hard drive can help the user mitigate the consequence of errors made when working with important files.

Example: Search for Human Remains of the Columbia *Crew*
Error-correction logic is a powerful tool that can be used to mitigate the consequence of errors already made. This is an example of error-correction logic applied by an Evidence Response Team (ERT) from the Federal Bureau of Investigation (FBI) to

identify and correct many errors in a large database of information used after the Space Shuttle *Columbia* accident.

For thirteen days in February 2003, more than 2,000 volunteers directly contributed to the search operations to find and recover the human remains of the *Columbia* crew with dignity, honor, and reverence. Of the various search methods used, the most effective was the line search. Teams of people walked in long lines, shoulder to shoulder, using Global Positioning System (GPS) navigation for tracking and documenting the locations of suspected human remains.

At the time, the field team personnel did not have much experience using hand-held GPS units. The field teams made four different categories of errors when they communicated the GPS coordinates to the search and recovery leadership team in the command center:

1. The most common error was transmitting the coordinates as degrees, minutes, and seconds, when the units should have been degrees and decimal fractions of degrees, or vice versa.
2. The next most common category of mistakes was making transcription errors by simply reporting incorrect digits.
3. Other errors occurred when different field operators were using handheld GPS units set to calculate coordinates based on different datum references, unknowingly.
4. Sometimes the teams transmitted stale coordinates displayed before the units locked on to the appropriate GPS satellites.

Had the analysis team in the command center not been able to correct these errors, the precise analysis would have been flawed, causing adverse effects on the search efforts.

As the operational search director, I will forever be thankful for the valuable assistance provided by the ERT from the FBI. This team, coordinated by Special Agent Mike Sutton, used an effective method to mitigate the consequence of erroneous reporting of GPS coordinates from the field teams by using error-correction logic. With so much experience in search-and-recovery efforts, the ERT personnel have learned to keep meticulous and comprehensive records of the reports received from the field teams.

Some of the recorded information was redundant, which allowed the error-correction logic to be applied. Because the ERT's database of the reports was so complete, the analysis team was able to correct each of the erroneous coordinates using logical deduction and intuition based on extensive experience, regardless of the type of error being committed.

With selfless devotion to the mission of finding and recovering the human remains of the *Columbia* crew, 2,000 volunteers demonstrated overwhelming and astounding dedication and professionalism.[28]

Example: Negative Pressure Test

In my career, I have discussed the best mental attitudes with some of the best drillers in the oil-and-gas industry. For three weeks, I spent time with one of them, a company man and former driller, Walter Guillot, from Louisiana. I was enthralled with his explanations, delivered in his thick Cajun accent, of his state of mind and how he thought when he occupied the driller's chair during his two-week hitches. I soon began to understand what it means to be either a driller or a company man (leader of the well-site operations) and what it takes to do both jobs right, with operating excellence.

To draw an analogy from the US Navy, the driller is the captain of the ship, and even has the chair (see photo 88). He or she makes the tactical decisions during drilling operations. The company man is the admiral of the fleet and makes the strategic decisions. He or she is the leader of the well site. Great well-site leaders and drillers are always watching, waiting for the well to try to kill them. As one of them explained, there is a "monster in the hole" in the form of hydrocarbons at 13,000 psi, five miles down. The great ones don't think about production or targets or revenue. Every thought is dominated by containing the monster. Every night, they go to sleep thinking about what they will do if trouble starts rising. A driller like Walter never turns his back on the well as long as he's in the chair and as long as his equipment is connected to the well.

Photo 88: Many decisions must be made during deepwater drilling operations.

---

28 Philip C. Stepaniak, *Loss of Signal* (Washington, DC: National Aeronautics and Space Administration, 2014), 18–31.

As if drilling for hydrocarbons is not hazardous enough, deepwater drilling presents additional challenges with the remoteness of the worksite under the sea and the higher pressures in the reservoir. Many decisions are made during deepwater drilling operations. Every one is important. The criticality of the decisions increases with the depth of the drill pipe.

As the well is drilled deeper and deeper during operations, the drill bit eventually breaches the zone of hydrocarbons trapped in the reservoir miles beneath the surface of the earth under a deep body of water. The hydrocarbons are squeezed tremendously under high pressure from the rock and seawater above the reservoir. Pictures of drilling rigs on land in the early twentieth century sometimes showed high-pressure blowouts that occurred when the drill bit breached the reservoir, resulting in spectacular—and dangerous—geysers of oil.

Modern-day drillers prevent these high-hazard blowouts (and wasted revenue) by pumping heavy synthetic drilling fluid down into the well during the drilling operations. The hydrostatic pressure from this heavy column of fluid overpowers the internal pressure in the reservoir and prevents the hydrocarbons from rising in the well bore until production of the oil is desired later.

Environmental regulations require the company drilling the well to remove the synthetic fluid after the well has been drilled. Before the fluid is removed, though, cement is pumped into the well. The cement becomes a temporary barrier that will prevent the hydrocarbons from escaping from the high-pressure reservoir into the lower-pressure well as the drilling fluid is removed. Later, when the company is ready to produce the oil, drillers will punch holes in the cement allowing the hydrocarbons to flow into the well and up to the surface.

Before the drilling fluid is removed, a final safety check must be performed to verify the cement has set properly and is in the proper position to seal the well from the hydrocarbon reservoir. This safety check, called the negative pressure test, is analogous to the pressure integrity test of the vestibule area between the hatches of the International Space Station and any visiting space vehicle (discussed in the technique Follow Procedures *Thoughtfully*).

Because only a small amount of drilling fluid is removed, the negative pressure test is a safe and controlled method for assessing the integrity of the cement barrier in the well. Under the slightly reduced pressure in the well, the hydrocarbons are actually encouraged to flow but only slightly and only if the cement has not sealed the reservoir properly. If the test passes, indicated by no flow out of the well, the cement barrier is assumed to be good, and the well should not flow after the hydrostatic pressure of the heavy drilling fluid is removed. If the test fails, the cement barrier is assumed to be ineffective, and the cement must be repaired before the drilling fluid is removed.

After a long introductory explanation for this example, I'm finally getting to the point of this subtechnique, Mitigate the Consequences of Errors. In drilling operations, the negative pressure test conducted before removing drilling fluid is critical as the operators check the integrity of the barrier intended to prevent blowout. In space operations, the vestibule pressure check before opening a hatch is critical as the operators verify integrity of the barrier to prevent loss of cabin atmosphere. Operators really don't want to make mistakes in either of these tests. But operators are human and humans make mistakes. What happens if either of these tests is conducted erroneously? Does an accident follow? Though it did on the *Deepwater Horizon* in the Gulf of Mexico on April 20, 2010, the consequences of similar errors can be mitigated.

In both deepwater drilling operations and deep-space operations, catastrophe doesn't necessarily follow after a mistake, even if the mistake was made during a critical test. In both cases, there is time for operators to Mitigate the Consequences of Errors, by taking the appropriate corrective actions. To do this, though, operators must be prepared, as Walter was during every minute of his two-week hitches, even when he was sleeping, as long as his equipment was connected to the well.

Operators must know what to look for in identifying errors and must know what to do if an error was made. Preparation enables success, but it's only half the battle. More important than knowing is doing. The operators must actually look for the error, must identify the error, and then must take the proper actions to mitigate the consequence and eliminate the danger. There is time to secure a well after it begins to flow, before a blowout. There is time to close a hatch before all the air escapes on a space station. This is the point of Mitigating the Consequence of Errors. We all make mistakes. Operators in hazardous environments must be skilled enough to correct their mistakes. Otherwise, they won't live long enough to make and correct more mistakes.

11. Develop and Execute a Plan (For All Critical Phases of Operations)

Success in complicated operations can only be achieved when operators execute a well-developed plan for all critical phases of the operations.

Developing a plan requires an understanding of the mission, the capabilities of the personnel who will be conducting the operations to accomplish the mission, and a deep understanding of the equipment and systems the personnel use. An overall strategy for accomplishing the mission must be known before developing the plan.

Before attempting to execute the plan, all personnel involved in the operation should study the plan and develop a thorough understanding of its contents. All personnel should train extensively before attempting to execute the plan. The plan should include provisions allowing the operators to employ the high-quality practices of operating excellence during execution, including the following:

- Maintain situational awareness during operations;
- Understand and protect limitations of the systems, equipment, software, and people;
- Maintain an outward focus and sense hazards immediately;
- Predict trigger steps, consequences, and "What if?" scenarios;
- Maintain eternal vigilance and expect failures during execution;
- Conduct after-action reviews following the various phases of the operation;
- Share lessons immediately;
- Assess progress of operations and be ready to modify the plan using proper analysis;
- Anticipate increasing risk and always be ready to invoke a contingency plan (see the following technique).

Example: Flight Plan on Space Shuttle Missions
Each flight plan we executed on Space Shuttle missions was developed over the course of a year with hundreds of meetings and planning sessions and hundreds of hours in simulations. An excerpt from our flight plan on STS-113 is shown in photo 89. Multiple deletions, additions, modifications, and edits were made, before launch and every day while on orbit.

When we flew the STS-63 mission in 1995, the best technology of the day for transmitting multiple changes to the flight plan via uplink was the teleprinter. Photo 90 shows Pilot Eileen Collins and Mission Specialist Vladimir Titov with the teleprinter changes for one day of the flight plan. In our $4 billion spaceship, we spent an hour each day with the changes, cutting and pasting with scissors and tape (as if I were back in the fifth grade dreaming about flying in space). By 2002, on my sixth mission, newer technology allowed us to receive scanned copies of the changed flight plan via e-mail.

When developing flight plans for spaceflight, the topic of many debates between the flight crew and the planners was how many activities should the crewmembers attempt. Most commanders did not want to oversubscribe their crewmembers during flight by scheduling too many activities in the plan. The common concern was the mission might be judged as a partial failure if the

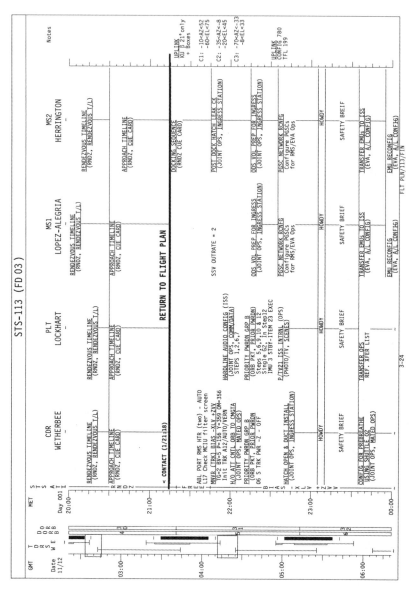

Photo 89: Excerpt from the flight plan for STS-113 on docking day showing scheduled activities for each crewmember

Photo 90: Pilot Eileen Collins and Mission Specialist Vladimir Titov, STS-63, with daily changes to the flight plan (Photo Credit: NASA)

crew did not complete all of the planned activities because they ran out of time. Additionally, if too many activities were scheduled, the crews risked becoming excessively fatigued, which might result in making more mistakes and increasing the danger.

On the five missions I planned and flew as a commander, I held a different opinion. I desired to be quite aggressive in planning and executing the missions. Because time on orbit was valuable, I attempted to schedule as many activities as I could fit into every minute of every day. I was willing to accept the possibility of accomplishing less than 100 percent of the scheduled activities. I was more concerned with taking all the time, effort, and risk to launch and then finding we were completing activities ahead of schedule and were wasting valuable time in space by having an insufficient number of activities scheduled.

So, I influenced the planners to develop a fully-loaded flight plan, and then I aggressively pushed the crewmembers during training to maximize our readiness to perform every activity on orbit. Our aggressive training, with skilled instructors, gave us the confidence and the ability to execute the full flight plan, while maintaining our high-performance standard of operating excellence.

With our aggressive operations, I inserted two "relief valves" to ensure safety. First, I made a preflight agreement with the flight directors that if I assessed we were getting overloaded on orbit, I would recommend deleting some of the secondary objectives. Because we made the effort to develop a trusting relationship, the flight directors always agreed. Second, our crew developed a fatigue-management strategy to help each other and operate safely when any crewmember assessed fatigue in any other crewmember. This strategy was developed specifically in conformance with the thirteenth Principle of Operations for spaceflight crews (refer to chapter 3). The techniques that were used in our fatigue-management strategy to improve the performance of potentially tired operators were based on some of the Techniques for Operating Excellence listed in this and the next chapter.

To summarize this example, I offer the following—and this list represents one of the most powerful lessons I have learned in hazardous operations:

- *High-quality Plan*: To be successful in hazardous operations, we must develop a high-quality plan.
- *Aggressive Plan*: To accomplish the most objectives possible, we must create an aggressive, high-quality plan. Many operators believe safety requires the conservatism of a nonaggressive plan. I disagree. An aggressive plan can be safe. Safety requires a deep understanding of, and complying with, safety requirements, safety philosophy, safety intent, and the standards and techniques of operating excellence. As long as each of these are infused in the plan and never reduced, we can build aggressiveness into our plan.
- *Training*: We must train aggressively with operating excellence to execute our plan.
- *Execution*: We must execute that plan while demonstrating the highest standards of operating excellence.

But here are the most important lessons:

- *High-Level of Awareness*: When executing our aggressive plan with operating excellence, we can be aggressive in our execution, but we must maintain a high level of awareness and continually assess our operations and anticipate when to modify our execution.
- *Modify Execution*: Modify the execution of our plan when we must. Modifying our execution at the correct time in the correct way is not a skill easily learned. We must always allow our "relief valves" to open; we must back off at the right time; we must invoke our contingency plan at the right time; we must be willing to sacrifice the mission and return safely to attempt another mission another time. When we do this well, we have achieved the elite level of operating excellence.

If we plan and execute the right way, we can operate aggressively *and* safely, simultaneously. Our team will accomplish more than anyone thought possible, and we will live to execute another mission another day.

Example: Rendezvous Plan

Photo 91 shows an overview of the final approach to docking with the International Space Station on STS-113. The entire rendezvous plan was two hundred pages in length and was a four-hour operation on orbit. The plan was the culmination of years of experience from experts in the agency that began with Jim McDivitt's rendezvous maneuvers on *Gemini 4* in 1965.

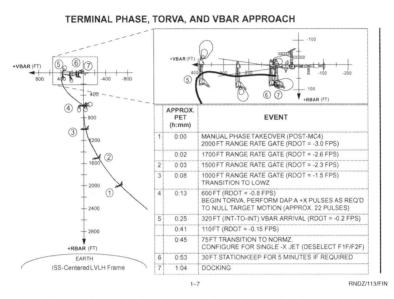

Photo 91: Excerpt from the Rendezvous Procedures for STS-113, showing an overview plan for the final approach to docking with the International Space Station

By the time we flew on STS-113, the engineers in the Loads Sections in the Shuttle and Station programs were becoming increasingly concerned with the docking performance on some previous flights—with relative closing speeds faster than the targeted value of one tenth of a foot per second. The concern was decreased fatigue life due to excessive stress on the docking hardware. Our task on the mission was to use training, experience, judgment, teamwork, and operating excellence to demonstrate how to execute the docking task flawlessly, and within acceptable performance limits of plus or minus three hundredths of a foot per second from the targeted value.

As the launch date approached, we knew from analyzing the flight control characteristics of the Space Shuttle why some previous missions hit the Space Station at higher velocities than expected, with higher loads than desired. We understood how to use the navigational tools that would give us situational awareness of our proximity and docking operations. We developed the priorities required to define mission success. We trained extensively with each other on the crew and with the flight controllers in Mission Control in Houston. We knew each other's strengths and weaknesses and how to pull together as a perfectly synchronized team. And we developed a plan that couldn't fail, with contingency plans if it did.

On November 23, 2002, we launched and executed the best plan in the best way possible, with the highest standards of operating excellence, because we had the best teams in flight and on the ground that day, made that way from the collective NASA experiences of rendezvous experts since 1965. The Shuttle and Station teams executed the plan flawlessly.

Example: Crew Briefings in Quarantine
A well-developed plan is essential for operators who are trying to accomplish complicated and dangerous missions. Typically, operational plans have sufficient details to cover the important operational aspects of the mission. What about the less important and mundane operational aspects, or the administrative details? Often, these are not included in an operational plan. These seemingly trivial details, if overlooked by operators, can destroy the operation, leading to failure of the mission.

As I gained experience in commanding space missions, I began to appreciate how each flight plan was developed with high-quality attention to detail in the operational aspects of our missions. Our operational training was superb, and we were well qualified to execute the flight plan in Earth orbit. But, some aspects of everyday life in space and even some operational aspects of the mission were not covered in any flight plan.

Before each mission I commanded, I decided to supplement the official flight plan with an addendum plan specifying our commitments to each other as a team and how we were going to operate together in space to be successful. Long ago, I learned groups of people working together could accomplish seemingly impossible missions. Their performance is always greater if they know exactly what is expected from them.

Each crew spends the week before launch isolated in health stabilization, formerly known as medical quarantine, to prevent acquiring germs that could degrade human performance on orbit. This was the perfect time to pull our crew together in a four-hour session, away from distractions, to set expectations and make personal commitments to each other. This was the basis of my addendum plan for mission success.

The following are some of the items we discussed. This excerpt is only intended to give a sense of the kinds of items normally not found in an operational plan. These items are just as important, though, if you and your team intend to succeed in a dangerous operation.

Operational
- Communication protocols used during the launch countdown
  - When to speak, when to listen
  - Including acceptable and unacceptable use of humor
- Communication protocols on orbit
  - How to be effective and efficient
  - How to sound professional
  - How to avoid being unprofessional
  - Being courteous to the flight controllers in mission control
- Commitment to listen to each other and operate with respect
- How to watch out for each other and offer assistance
- Times in the flight plan when we can get ahead in the schedule
- Times in the flight plan when we likely will fall behind
  - How to catch up
- Physical and mental challenges of operating in zero g
  - How to move
  - How to protect equipment
- Honoring sleep periods
- How to strive for perfection
- How to operate with courtesy
- How we intend to operate if a disaster is impending
- Physiology during reentry
- Mental techniques for operating under physical and psychological stress

Administrative
- Culture and traditions on the International Space Station
- Welcoming the expedition crew in space
- Change-of-command ceremony
- Honoring personal space in space
- Use of e-mail, including precautions, courtesy, ethics
- Food preparation techniques and efficiencies
- Wet and dry trash management
- Keeping track of personal gear; lost-and-found policy

- Efficient ways to change clothes and conduct personal hygiene activities in zero g
- In-cabin video cameras
  o   Always assume video is on
  o   Always assume audio recording is on
  o   Be aware of camera location
- Personal hygiene responsibilities in a closed environment
- Expectations from our guests on the International Space Station

12. Have a Contingency Plan

In a dangerous business, operations are volatile, uncertain, complex, and ambiguous.[29] To accommodate necessary changes because of uncertainties, failures, and the normally dynamic environment, the original operating plan should have robustness and redundancy built in. Failures will occur and changes will be needed. No matter how good the original plan is, it might not be good enough when failures accumulate in number and compound in danger. When too many failures occur, especially when they combine in dangerous ways, it will be time to abandon the original plan and invoke a contingency plan.

Always *Have a Contingency Plan* ready to accommodate problems encountered while executing the primary plan. Have the ability to recover from any degradation. During flight, pilots maintain a divert plan in case the weather deteriorates. They never arrive at a runway with insufficient fuel to fly to another airport. Even in good weather, another plane may crash on the runway in front of their airplane. Pilots are always prepared to land not at the primary destination but, rather, divert to an alternate airport.

Example: STS-63 Leaking Jet
At 12:22 a.m. on February 3, 1995, the crew of STS-63 launched aboard the Space Shuttle, *Discovery*. Our primary mission was to test and evaluate the rendezvous and proximity operations of the Space Shuttle with the Russian Space Station, *Mir*, in preparation for the first docking mission scheduled to occur three months later. Our goal was to follow the procedures, use the systems, and manually fly *Discovery* to a relative distance of ten meters away from *Mir*.

I had the privilege of flying with Colonel Eileen Collins, US Air Force, the first woman pilot astronaut at NASA. With cool professionalism that is her signature, she superbly handled multiple malfunctions in *Discovery's* propulsion system

---

29   I had the honor of meeting and learning from a superb leader, US Army Colonel Bernard B. Banks, PhD, professor and department head, Behavioral Sciences and Leadership, United States Military Academy at West Point. He uses the acronym VUCA when he teaches future Army leaders to anticipate that operations often will be volatile, uncertain, complex, and ambiguous.

during her first launch and ascent to space, from 0 to 17,500 mph in eight and a half minutes.

Soon after we were on orbit, we could see our primary mission was in jeopardy. Our normally spectacular view of Earth and space was marred by an equally spectacular leak of oxidizer from one of our Reaction Control System jets that had malfunctioned during powered flight. Eileen performed without error to put us in a safe configuration, but valves in the jet couldn't be sealed. The leak was resulting in frozen oxidizer particles being released in a cone-shaped pattern extending several miles into the void of space. Unless we could stop the leak, we would not be allowed to rendezvous with *Mir* two days later. None of us were willing to risk the lives of the cosmonauts by possibly contaminating the optical sensors on their *Soyuz* capsule used for reentry to Earth.

One of the strengths of NASA engineers is their ability to create plans, redundant plans, and contingency plans for the day when the original plans begin to crumble under the weight of failures upon failures. We always expect failures (and we're rarely disappointed in these expectations). In Moscow, American flight director Bill Reeves coordinated the contingency plan with the Russian and American ground control teams.

On Flight Day 2, the task fell to the world's fastest and highest pilot, Eileen, to work the contingency plan and attempt to stop the leak of our malfunctioning jet. She is shown in photo 92. This plan was intricate and extensive—actually, desirable features when working with technically complicated systems in space. Eileen executed the procedure flawlessly, typical of her performance during the remainder of this flight and her three future spaceflights. Together, the flight team in space and the ground teams in Houston and Moscow executed the contingency plan successfully. Unfortunately, although the leak was significantly reduced, the jet was still leaking a small amount.

Photo 92. Pilot Eileen Collins, STS-63, executing a contingency plan to reduce the amount of propellant leaking from the Space Shuttle, *Discovery*, and prevent contamination of the critical safety systems on the Russian Space Station, *Mir*. (Photo Credit: NASA)

While Eileen worked the contingency plan, the rest of us continued with preparations for the rendezvous operations. At more than five miles per second, *Discovery* was closing slowly (in a relative sense) on *Mir*. The next day would be decision day. As we went to sleep that night, I told my crewmate, Vladimir Titov, the Russian cosmonaut flying with me as a mission specialist on board *Discovery*, that if the leak didn't decrease further, I wouldn't risk the lives of his fellow cosmonauts on *Mir*.

In a testament to the value of having contingency plans—and the skills of our pilot, Col. Collins, and the teams in the Mission Control Centers—we woke up on rendezvous day with gorgeous weather. The snowstorm from the jet had subsided, and the day was sunny and clear with unlimited visibility and extremely light winds (even at 17,500 mph, there aren't many molecules blowing by). The flight crews of *Discovery* and *Mir*, with the ground teams in the US and Russian space programs, completed the joint mission successfully by conducting the close approach to ten meters. Having a contingency plan ready and executing that plan when needed saved the mission of STS-63.

There is one important caveat to having a contingency plan. If problems are encountered while attempting to execute the primary plan and you don't have a contingency plan, don't attempt to create a new contingency plan on the fly. If you have not already taken the time to develop a well-thought-out contingency plan, it's too late. You may overlook a critical issue, and you may actually increase the danger to operating personnel if you try to create a plan in the heat of battle. At that late stage, your best option when the primary plan fails is to put your team and your system in a safe condition. Regroup, with operations suspended, and begin to create thoughtful alternate plans.

Example: Diamond-of-Diamonds Flyby During an Airshow
This is the story of an operation that became unsafe when an inadequate contingency plan was developed and executed on the fly, without proper thinking and analysis.

Naval aviators are trained to fly in tight formation for two reasons. Naval air warfare strategy is enhanced with tactical formations of multiple aircraft fighting as teams, and aircraft carriers are less vulnerable with expeditious recovery of aircraft arriving in tight formations. To maintain proficiency, naval aviators fly in formation on nearly every flight. Photo 93 shows a ten-plane formation flight I documented for training over Jacksonville, Florida.

In 1980, I participated in an airshow at sea for dignitaries who were visiting our aircraft carrier. The final element of our airshow was an impressive, sixteen-plane

Photo 93. Naval aviators from VA-72 maintaining proficiency in formation flying over Jacksonville, Florida

formation, "diamond-of-diamonds" flyby. A lead diamond of four F-14 Tomcats, always a crowd-pleasing favorite, headed the formation. In the trailing diamond of four A-7 Corsairs, I was flying on the right wing. Our executive officer was in the slot position, aft and left from me, so he was directly behind our lead A-7 and flying as the tail in the center of the whole formation of sixteen airplanes.

Our trailing diamond formation of A-7s was stepped down about fifty feet in altitude from the lead F-14s. The lower altitude helped us avoid the turbulent exhaust from the twin, afterburning TF-30 turbofan engines of the "turkeys" (the attack pilots' derogatory term for the Tomcats, whose fighter pilots thought they occupied the top echelon in the culture wars of naval aviation). Two diamond formations of A-6 Intruder and S-3 Viking aircraft were on either side of the large formation.

As a junior officer in the air wing, I kept my mouth shut and my ears and eyes open observing. I was equally as impressed with the preflight briefing as I was with the actual formation flight. Both were conducted with precision. For the briefing, the senior officer, who commanded the entire airshow demonstration, stood in front of sixty crewmembers in the ready room of one of the fighter squadrons. Safety and operations are indistinguishable when flying around the ship; but, with sixteen airplanes, flying next to each other, at low altitude, safety took on a higher meaning. If any pilot didn't listen and pay attention to the rules in this briefing, he would be dead. Maybe literally. And if so, he would likely take others with him in mangled airplanes to the bottom of the sea.

The plan was necessarily detailed. We would rendezvous at three thousand feet, with sufficient clearance above the water. The flight leader's control of the formation would be precise. The commander in the lead F-14 would slowly and deliberately descend to five hundred feet, and no lower, for the precision flyby past the ship. The commander stressed he would descend no lower than five hundred feet because the A-7s were even lower in the trailing diamond.

On the flight deck, a coordinated team of deckhands, plane captains, flight crew, and ship's company officers began the intricate tango that is aircraft carrier launch operations. At the precise time, the first of thirty airplanes thundered into the air, shot off the ship by a steam catapult. Twenty minutes later, when all the noise and activity subsided, twenty-nine airplanes were in the air. One F-14 remained on the flight deck, having gone down for a maintenance issue.

Airborne, the commander reconfigured the sixteen-minus-one-plane formation, as planned in the contingency operation. The slot F-14 (in the rear of the lead diamond) was ordered to fill the hole on the side of the diamond created by the missing F-14. This meant that the new hole was where the slot F-14 used to be, leaving the lead group of three F-14s symmetrical again, as they led the modified fifteen-plane formation, with us still in the rear diamond.

The big diamond-of-diamonds flyby was the finale of the show. Before our pass, other demonstrations were scheduled. One element was the always exciting, opposing pass of two high-speed F-14s at low altitude, in full afterburner.

This is where our problems began. After the opposing F-14s successfully completed their pass, one of those fighter pilots decided he was going to fill in the hole in our incomplete diamond-of-diamonds. Making his error doubly worse, he chose not to tell anyone on the radio about his plan with a flaw. Tripling down on his egregious error, making it off-scale bad, he began to join our formation at low altitude after we had already completed the rendezvous up at three thousand feet— approaching the ship at five hundred feet, fifteen pilots flying as one, throttles and flight controls balanced with precision.

What I saw next was the surprise of my short flying career and would have been a beautiful sight had I not been sucked into the vortex. Rising slowly from below, attempting to fill in the hole directly in front of us, was a mesmerizing, spectacular, and deadly view into the hot end of the turkey's engines. In the self-absorbed world in which fighter pilots play, he must have forgotten that, to ease up into the slot position, he first had to blowtorch the insignificant A-7s in the face. At least, I hope he forgot and wasn't evil on purpose. I imagine his radar intercept officer sat mute in the backseat, clueless and innocent.

The four of us in the slot diamond were instantly blown into slow-motion hell. On the right side of the formation, turbulence from the turkey's wingtip vortices and engine exhaust was sucking my airplane uncontrollably toward the left, my port wingtip headed straight for my XO's helmet in the tail of our diamond. The closer I got, the more right lateral stick I applied, but it wasn't working. Inches from breaking his canopy and his faceplate with my wingtip, my final option was to step on right rudder. With a sharp audible bang from the chaotic airflow, this worked and I broke free from the invisible force sucking me in. My XO felt the bang and thought I had a midair collision with his airplane. I didn't think so, but that was irrelevant. Pilots don't try to land an airplane on an aircraft carrier based on merely thinking the airplane is undamaged.

The whole formation broke apart, without conducting the flyby, and immediately went to a safe configuration, far away from the fighter pilot who sent us to aviators' hell. With two wingmen performing safety chase, my XO and I separately conducted our emergency procedures to assess flight control and handling qualities in the low-speed landing configuration. Better to learn the airplane is unflyable at a safe altitude rather than on short final behind 80,000 tons of steel ship in the middle of the sea. Luckily, both airplanes were controllable. After successful flight control checks, every pilot landed aboard the ship in tight sequence. Postflight inspections revealed our airplanes had made no contact with each other.

As soon as my right rudder worked and I broke free from the jet wash, my troubles were over. The offending fighter pilot's troubles were only beginning. After safely landing, all the pilots reconvened in the ready room. This time, a majority of the air wing seemed to be present to witness the firestorm.

With his eyes in afterburner, our commander began to berate the offending fighter pilot. "Don't you EVER join a formation without clearance! Don't you EVER join a formation UNANNOUNCED!! Don't you ever JOIN A FORMATION BELOW 500 FEET!!! DON'T YOU EVER **CHANGE A PLAN THAT WE BRIEFED!!!!** *DON'T YOU EVER TRY TO KILL MY PILOTS!!!!!"* The humiliation continued for some time. After a while, I began to feel sorry for the pilot, even though he was a fighter pilot. Creating thoughtful plans wasn't really his strong suit.

We learned an important lesson that day in Carrier Air Wing One. Always have a contingency plan. But develop the plan ahead of time. When things go badly, don't create a new plan on the fly. Don't send the operations to hell. Go to a safe condition and live to plan and operate another day.

13. Preserve Options During Operations

Some of these thirty techniques are intended to help operators identify the subtle signals of the next accident. But, identifying the signals is only part of the solution. To save the mission, an operator must be able to prevent the accident. And the accident can be prevented only if the operator has options available.

This technique, *Preserve Options During Operations,* is intended to influence the operator to maintain at least one final option to prevent or avoid an accident. If subtle signals of an impending accident are identified, and the operator has no options, nowhere to go, and has "painted himself into a corner," he can only hope the final failure won't occur to create the accident. In a dangerous business, we don't rely on hope.

Accidents and injuries can be avoided, but only if options are available. Don't get trapped without an escape route. While driving, don't enter another driver's blind zone without an option or contingency plan to escape immediately, should that vehicle begin to veer into your lane. While driving on a stormy night, if you hit a deer after a blind curve, the mistake was not in hitting the deer; the mistake was driving at a speed high enough to eliminate the final option to stop or avoid the impact.

Predict failures and deteriorating situations. Don't get trapped. Never relinquish your last option.

Example: "The Box"

At one point early in my aviation career, our safety officer asked me to give a presentation at our next safety meeting. As a junior officer in the squadron on my first tour of duty aboard an aircraft carrier, I had little flying experience compared with the other twenty A-7 pilots. The pressure to stand in front of hyperconfident pilots was actually greater than the stress of a night carrier landing in bad weather, so I wanted to create something useful to discuss. The topics in previous safety presentations were varied but seemed to be always narrow, fact-based, objective, and fairly uninteresting. If I was about to be shot down in flames by twenty naval aviators, at least I wanted to make my presentation thought provoking, and give them something that might save their lives, including one cheesy visual aid to help them remember the concept.

"The Box" was a term we used to denote a deteriorating operational situation with limited options for escape. The walls close in, crushing your ability to think clearly, and squeeze the last drop of confidence from your psyche. You never wanted to be put in the box.

One of the other junior officers spent a particularly harrowing night in the box, trying to land aboard the carrier. The weather was horrible, the visibility was worse, and he couldn't see the small, dark pitching deck in the bigger blackness. Three times he boltered, as his tailhook missed all four arresting cables. Each time, he had to psych himself up impossibly yet again for one more landing attempt, the next one scarier than the previous. On his "trick-or-treat" pass, meaning only enough fuel for one final attempt, somehow he landed. As he taxied clear of the landing area, he shook, still sitting in his pool of sweat, destroyed confidence, and unused ejection seat. His last option wasn't needed that night, though it was close.

I started thinking about the box and how to not be put into one. That's what I talked about at the safety meeting. Diagram 4 shows the box I made to explain the concept. The four sides can close tightly, removing escape options and giving you a bad day as an operator. In any given operation, part of your system may succumb to a failure. That's one side, slammed shut. The environment in which you are operating may suddenly deteriorate. Door number 2, locked. There may be a latent deficiency in the design of your system. Door 3, throw away the key. Finally, you may make a mistake. If all four of these challenges occur at the same time, you may find yourself boxed in, with no way out. So, your goal was to somehow prevent all four sides from slamming shut.

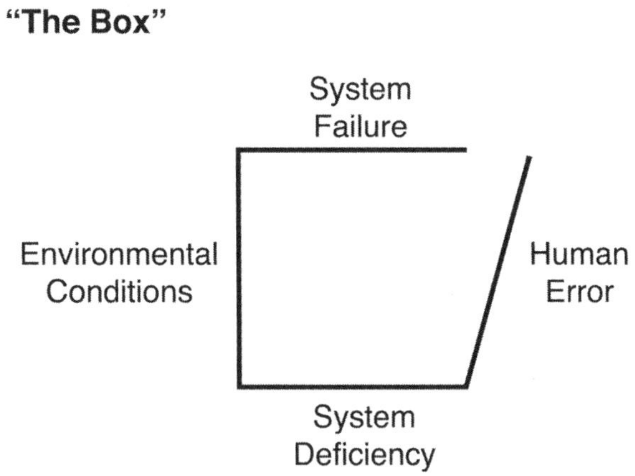

"The Box"

Diagram 4: "The Box." During dangerous operations, don't get put in the box.

But, preventing all four constraints from happening is untenable. First, on my initial cruise, the ship's navigator often found horrendous weather to sail through, right when we were trying to land aboard the ship, so we often had environmental barriers blocking our safety valve. Second, the A-7 had numerous design deficiencies, including a single turbofan engine, with extremely slow spool-up time, especially when we needed immediate power during final approach to landing. The deficiencies designed into our systems were always present. Third, the pilots couldn't reliably prevent failures from happening in their airplanes. On many flights we experienced at least one system failure. And, finally, I had begun to realize human error is inevitable in human pilots. In no time at all, we could find ourselves boxed in by four independent constraints leaving us with no options to avoid a potential accident.

There had to be something else we could do. Diagram 5 is an illustration of the cheap cardboard visual aid I made to help the pilots remember the concept in three dimensions. The sides of the box represented the same four operating conditions that can reduce or limit the options for the pilot. But now our expanded 3D box of options, which better represents the operational situation, has two extra sides.

I faced bad situations before and was saved by luck. So, as I explained, the bottom of the box represented luck. Remember, though, I can't rely on luck to be there when I need it, just as I could not control the other four sides of the box. The

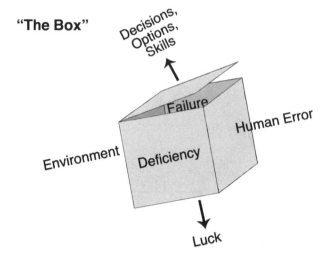

"The Box"

Decisions,
Options,
Skills

Failure

Human Error

Environment

Deficiency

Luck

Diagram 5: Don't make bad decisions, run out of options, or fail to develop your skills in dangerous operations. If you do, "the Box" will close and leave you no way out.

key to operating excellence is keeping the top of the box opened. This I can control. As an operator I can perform in ways to keep the top of the box open, which gives me the ability to escape at any time from any deteriorating operational situation. I train to develop the proper *skills*. I employ professional discipline to preserve *options*. I use good judgment, based on experience and high values, to make the right *decisions*.

Learning to keep the top of the box open is why we train so intensely and extensively in hazardous operations. As pilots, we study our aircraft systems and develop deep technical knowledge of how the systems work. We train correctly so we develop the proper skills. We can't eliminate errors, but we can develop the skills to recognize errors immediately, take corrective actions, and reduce the consequences.

When we do experience system failures, we are able to combine our technical knowledge with skills developed through practice in the simulators to be able to handle any emergency effectively. When the weather begins to deteriorate, we make the right decisions and preserve options. If we are caught in weather that has become unexpectedly worse than the operating limits allow, we have honed our skills to such a high level of proficiency we are able to fly safely and confidently in any weather conditions.

We have the professional discipline to use the systems properly with operating excellence. We know the limits of the systems and our personal limits. We have the wisdom to push ourselves aggressively yet always within the rules, which allows us to accomplish more missions with higher quality results. We can't avoid being put in the box, but we know how to keep the top open and preserve options during operations.

Example: Riding a Bike in the City

Here is an example that shows how to preserve options on a road bike in the city. The transportation code in my home state specifies a person operating a bicycle "shall ride as near as practicable to the right curb or edge of the roadway"[30] unless a parked vehicle prevents the person from riding safely next to the right curb. I intentionally created the situation depicted in photos 94 and 95 to show what can happen if I ride close to a parked car I am passing. Depending upon my speed, I may not have the option to avoid a door on the driver's side that is suddenly opened just before I pass. Almost instantly, this situation can become fatal.

Photo 94: The rider (me) has limited or no option if the door on the driver's side is suddenly opened.

Photo 95: I might have insufficient time to make an evasive maneuver.

To preserve options when passing parked vehicles, I simply ride as shown in photo 96, anticipating the door of the parked car may suddenly open. Simultaneously, though, I must preserve options to escape a potential hazard from behind, so I

30   Transportation Code; Title 7. Vehicles and Traffic; Subtitle C. Rules of the Road; Chapter 551. Operation of Bicycles, Mopeds, and Play Vehicles; http://www.statutes.legis.state.tx.us/docs/tn/htm/tn.551.htm.

maneuver to the left only after verifying I will have safe clearance from cars behind me that may be passing on my left side.

Photo 96: Preserve options by riding a safe distance from the parked car.

Example: C-130 at Low Altitude

Look at photo 97. Do you think the pilot in command was using a technique to preserve options during this low-altitude pass over the drop zone? What options remain at such low altitude if the flight crew experiences an engine failure, or is momentarily distracted, or has degraded flight controls with hydraulic problems, or is surprised with a runaway trim actuator? Don't get caught in your operations with so many failure modes and so few options to escape if a failure occurs.

Photo 97: US Army C-130 during low-altitude flyby. Does the pilot have a way out if an engine suddenly fails?

Example: Aerial Refueling During Flight Operations at Sea
I learned the technique of preserving options during operations in a stressful way, over two terribly uncomfortable hours. One of our missions in the A-7 was to transfer fuel to other naval aircraft in flight. Usually, on these in-flight refueling missions, we gave fuel to F-14 Tomcats at the beginning of their flight cycle to allow them to complete their training missions in aerial dogfighting in full afterburner.

Before transferring fuel to each airplane, the *giving* pilot tells the *receiving* pilot how much fuel to expect. Usually the fighter pilots asked for more fuel than I was willing to give. Some fighter pilots didn't seem to consider their fuel quantity gauge would never indicate the exact amount I gave because they had two thirsty engines burning fuel while their tanks drank from mine. Early in my career, when I was inexperienced and trying to be helpful, I gave them as much fuel as I could. On one flight, the ship experienced a delay in commencing the landing operations and that's when my challenge began. During the next two hours, I had to concentrate intensely and fly as smoothly as I was able at the exact airspeed to maximize my endurance and hope the ship wouldn't experience any further delays in getting me aboard before I flamed out.

For every minute of those two hours, I regretted giving the fighters as much fuel as I could possibly have given. I surmised few fighter pilots appreciated the difference between my big sacrifice of fuel and a slightly smaller sacrifice.

To preserve options during operations, and to protect myself if there were any delays in landing aboard the ship, I began to give less fuel on subsequent flights. I relaxed much more while waiting to land. The fighter pilots probably lost only one extra minute of training, while turning and burning in full blower doing their fighter pilot thing. Regardless, they still complained I hadn't given enough fuel. I tried to sound genuinely sorry on the radio, while I smiled inside my oxygen mask knowing I made the right decision to preserve my options during operations.

Example: Clearance Behind Automobile At Stop Light
I offer one final example to bring the concept of preserving options into everyday operations. Notice photo 98. I am stopped at a traffic light behind another vehicle. In this position, I have preserved my options by leaving sufficient clearance between our vehicles. If I need to move for any reason, I have sufficient space ahead to maneuver my vehicle. I am less vulnerable to being rear-ended by a distracted driver behind me because I have space to move forward. If I fail to anticipate a collision and I am hit from behind, I might be able to prevent a chain reaction by not hitting the vehicle in front of me.

Because I always plan to leave sufficient clearance ahead, I decelerate more cautiously and stop earlier. With this reduced deceleration, I might be avoiding a sudden skid in undetected slippery conditions. If I experience braking problems, or my foot slips from the brake pedal, I may be able to steer out of the way or activate my emergency brakes before hitting the vehicle in front of me.

The simple act of leaving space in front of my vehicle allows me to preserve options on the rare day I may need one.

Photo 98: Preserving options while stopped at a traffic light

14. Reduce Exposure to Hazards

By far, this technique has the most power to reduce injuries to personnel. An injury can't occur if the operator is not exposed to a hazard. And when the technique is used to protect systems by not exposing the equipment to a hazard, damage to systems can be prevented.

Even if a manager or operator decides the operation is safe enough, that is, the risk is calculated to be less than the allowable limit, this technique, *Reduce Exposure to Hazards,* should still be used to increase the margin from danger. The further you operate from danger, the more likely will be your chances to complete the mission—and live to fight another day.

Increase your distance from danger by eliminating or reducing your exposure to hazards. This single technique can be used to prevent injuries and accidents more than any other technique. Whether the risk is high or low, or the likelihood of injury is great or small, reducing exposure to hazards will reduce the number of painful outcomes. Don't rely solely on chance or the low probability of being injured to stay healthy. Avoid the hazard to avoid the pain.

You likely encounter multiple opportunities to reduce your exposure to hazards every day. One example is washing your hands before eating in a restaurant or in field operations. This takes extra time and effort, but your exposure to germs will be reduced with a proper process of washing.

On a rainy day, you might consider jumping over a puddle to avoid getting your shoes wet. Depending upon your skill at jumping, the probability you may slip and fall after landing may be low, though it is non-zero. If you take a few extra seconds to walk the long way around the puddle, you can't slip after a landing from a jump you didn't take.

Another example is wearing personal protective equipment. I always wear my form-fit musician's earplugs when I play drums to reduce my exposure to the obvious hazard of noise. But, these days, I even wear my earplugs at the movies. Though receiving hearing damage in one sitting is unlikely, I can reduce the likelihood to almost nothing by taking a few seconds to insert my earplugs.

Now consider different levels of risk in operations. In low-risk operations, sufficiently wide margins of safety from hazards exist, so *constant* vigilance is not necessary to maintain a safe condition. If conditions degrade slightly and the risk increases a little, the safety margins are reduced but only by a small percentage. People who work with low risk are not exposed to great hazards and don't need to conduct continual assessments of risk or modify their decisions or strategies to remain relatively safe.

To illustrate relative safety in different operations, Wayne Hale, NASA flight director and Space Shuttle program manager, once made an interesting comparison of the factors of safety for elevators, airplanes, and Space Shuttles. The elevator cables in an office building have a factor of safety somewhere near twelve. These cables can withstand twelve times the placarded weight before failing. Inspections are required but can be infrequent. Additional back-up systems, such as fail-safe brakes, are in place.

In medium-risk operations, the margins of safety are narrower. Airplanes have a factor of safety of 1.5, so inspections are more frequent and rigorous than in low-risk operations. Because these safety factors are lower, with tighter margins from real danger, more frequent assessments are required to maintain safety. Reducing exposure becomes more important. To provide a constant reminder of hazards, I think of margins of safety as *margins from danger*.

Operators who work in high-risk operations are perilously close to the edge of danger. When operating conditions change, even if only a minor change, the safety margins can become quite small, and tiny unseen perturbations can cause large failures. In some parts of the Space Shuttle, the factor of safety is 1.4. In other parts, it is as low as 1.1. With a factor of safety that low, the best way to preserve safe outcomes is to reduce exposure to hazards. To stay alive, operators must consider adjusting their process or environment to reduce exposure to hazards.

In low Earth orbit, space-walking astronauts can reduce their exposure to suit damage from incoming micro-meteoroid/orbital debris by choosing a route behind the protective structure of the Space Station when transiting from one work location to the next. The route may be longer, and transiting behind the structure may take extra time and effort, but the suit won't be hit. On the station's front side, the astronaut will never see a hypervelocity, killer particle coming. On the back side, the particle can't see the astronaut coming.

Good drivers use wise control of risk by reducing exposure to hazards when operating motor vehicles on the roads. The best drivers will not experience collisions on highways because they operate their vehicles with nearly constant attention to the dangers around them. They drive in a way to maximize their margins from danger by reducing their exposure to hazards. They don't put their car in a position where another driver can hit it.

Independent of the level of risk, the most effective way to prevent accidents is to reduce exposure to hazards. Regardless of the likelihood of injury, if you eliminate your exposure to a hazard, that hazard can't hurt you.

Example: Boat Officer Duties in Barcelona

These next two examples serve dual purposes. In addition to illustrating the importance of reducing exposure to hazards, each example contains leadership lessons I learned in trials-by-fire as a young Lieutenant Junior Grade in the US Navy.

The art of leadership is learned through experience and mentoring over an entire career. Though anyone can learn about leadership in the classroom, leaders are made in the field. The Navy excels at throwing junior officers into the breach on the high seas. Fewer officers are commissioned in the Navy than in the Air Force or the Army, so naval officers are given leadership experiences earlier in their careers.

One of the collateral duties of a junior officer in the Navy is to serve as a boat officer on the transport vessel used when the ship is anchored in the harbor. To transfer the sailors and Marines between the ship and the port city for well-deserved liberty after several weeks at sea, the Navy uses boats with a capacity of more than a hundred personnel, sober or drunk. The boat officer serves as the officer-in-charge on the small naval vessel, even when higher-ranking officers are aboard as passengers. The junior boat officer usually has little leadership experience. The boat is small, when compared with an aircraft carrier. The responsibility is not, as I was about to learn on my first cruise in 1979.

A few days before the ship pulled into harbor for our first port call in the Mediterranean, the junior officers in the Air Wing were summoned to the wardroom for mandatory training. A black-shoe (ship's company officer) announced he was commencing the training required to qualify us as boat officers.

The lights dimmed and a sixteen-millimeter projector began to roll a film. With dramatic music and somber narration, a Navy-produced movie recreated a tragedy that occurred two years earlier in Barcelona's harbor. On January 17, 1977, more than ninety Naval and Marine personnel were returning to their two ships after liberty ashore. Just after rounding the jetty in rough seas with high winds and low visibility, the Navy boat collided with a merchant ship and was capsized. Forty-nine people drowned trying to make way to the safety of their home away from home, anchored in the harbor.

The short movie ended and the lights came on. The officer stood and completed the training by saying, essentially, "As the boat officer responsible for the lives of your people, don't let this happen to you." Through watching the movie and being present to hear that one sentence, we had just become qualified boat officers. In hindsight, I give the officer credit. He wasn't being lazy by keeping his comments short. In one sentence, he elegantly summarized the brutal responsibility. Command-at-sea makes no allowances for limited experience or junior rank. Command-at-sea only requires completing the mission and protecting the crew and vessel.

Photo 99 is a picture I took of the jetty in the Barcelona harbor, around the time I began my duties as boat officer.

Photo 99: Boat officer duty; jetty in harbor at Barcelona, Spain

On my third trip on the transfer boat, during a short span of about ten seconds, I learned what it means to be a leader. At about three o'clock in the morning, with a hundred passengers, my three-person crew had just tied off at the base of the ladder to the aircraft carrier. The massive ship and the small boat were responding much differently to the dynamics of the violent sea state. In high winds and four-foot swells, the smaller boat kept rocking and rolling, heaving and smashing, into 80,000 tons of stable ship.

On board, the people situation grew more dynamic than the sea state. The sailors and Marines—mostly young, mostly strong, and mostly drunk—saw the stability and the warmth of their floating home only a few feet away. The collective mass of ex-high school football players began to fight its way to the head of the line, as each person tried to escape impending nausea. In an instant, the thought occurred to me they faced a hazard far greater than the high likelihood many of them would soon be sick, bruised, and bloodied. To disembark, and stay alive, each person had to time a single jump successfully, from our moving vessel to a stable platform. If mistimed, that single jump would be their last as they were swallowed into the blackness and crushed between the vessel and the platform.

I sensed I had only a few seconds to say something before someone died under my command. I always understood leadership is not the rank you wear on the uniform. Command authority is given to you, not by the chain of command above you but by your followers, and only if they choose to follow you. So, I searched for the words to inspire my followers to restore order.

Nothing came to me. I had no idea what to say. At that moment, I was certain of only one thing. I had to eliminate their exposure to the hazard. Without saying a word, because my mind was blank anyway, I jumped into the space between the handrails on our boat and gripped each side tightly to block their exit. I was a speechless and inexperienced officer. But, no one was going to die on my watch that night.

Suddenly, a most amazing thing happened. The drunken mob began to police itself. Someone shouted, "Hey, clean up your act; he ain't letting us off until we straighten up!" As quickly as the chaos started, orderliness returned. I stepped aside but stayed close to the front to grab anyone who might have slipped. Under the rough conditions, I don't think anyone saw my shivering.

I learned two things on that cold, dark morning in the harbor. When you don't know how to inspire your people, at least try to protect them from harm. One of the better ways to protect your people is to reduce their exposure to hazards.

Example: Boat Officer Duties, Lesson Two

Having survived the experience in the previous example in my small crucible of leadership, I felt a little more confident as a boat officer. A few months later, I was serving again as the duty officer on a transfer boat from the USS *John F Kennedy*, which was anchored just off the Port of Naples, Italy, as shown in photo 100. I took this picture a few days before I learned this second lesson about leadership and reducing exposure to hazards.

One afternoon in poor visibility under foggy conditions, I was shuttling sailors and Marines to Naples at the start of their liberty period. During a return trip to the

Photo 100: Boat officer duty; on the transfer boat from ship to shore at Naples, Italy

ship, I had a coxswain, who was driving the boat, one crewmember functioning as a deckhand, and no passengers. The three of us rounded the jetty and headed toward the ship, barely visible through the saturated air.

About halfway to our destination in the harbor, the visibility began to worsen, the ship disappeared, and my driver decreased our forward speed. In the thick fog, I could see the driver was becoming confused about which direction to steer. Suddenly, his mental model of our location relative to the ship became clear, and he throttled up again to high speed. Unfortunately, his mental model, though clear, was also wrong. Instead of steering toward the ship, he was on a bearing headed straight for the rocky jetty. I ordered him to come starboard to a new heading that I knew to be on the correct bearing for the ship. He ignored my order and seemed to accelerate. I gave him one more chance to comply by repeating my order. Again, he disobeyed and I knew what was happening. He thought I was trying to kill him. He was not challenging my authority from a place of insubordination; he was running from a place of fear. My order, which he believed without doubt to be wrong, only heightened his fear. His increased stress level solidified his incorrect mental model, and he tried to run faster from what he thought was certain death on the rocks.

At that brief intersection in our lives, where we had not yet developed mutual trust, I had a simple choice, with only one viable option. I calmly asked him to stop the boat. He immediately complied, knowing we couldn't hit the jetty if the boat wasn't moving. I watched him in silence for a full minute, allowing him to calm down. Eventually, using the most peaceful voice I could muster, I said to him, "One of two things is going to happen. Either, we will continue to sit here quietly, and not move—and that's okay with me; or, if you decide to move the boat, we will go in *my* direction, as slowly as you want. I will not give you an order. It makes no difference to me which option you choose. We will sit here all afternoon, if you want. But, we are not going in *your* direction."

There we floated, motionless and silent, for a long time. He never asked me why I thought the ship was in the direction I indicated. I never asked him why he didn't believe his compass because his brain was fighting a problem that did not have a rational solution. Pilots know the mind is easily confused when familiar references are stripped away. In his external—and internal—fog, his brain was desperate for meaning. Just before I gave the order to turn toward the ship, he thought he had visually latched onto the beacon of the lighthouse at the end of the rocky jetty in the direction I wanted him to go. Outside his mind in the real world, the rotating beam was bouncing around in the mist, giving him misleading indications. Inside his mind, he was convinced I was steering him into destruction.

I could tell he never did believe me about our correct orientation. But, eventually, he decided to move in the direction I wanted, as slowly as he wanted. After many long minutes, as he scanned quickly and breathed slowly, the giant aircraft carrier suddenly appeared out of the fog dead ahead. He turned and looked at me with the slightest of smiles and didn't say a word.

Leadership is not about giving orders. In a dangerous business, leadership is sometimes about reducing exposure to hazards, perceived and real. I had only one option to reduce the exposure to his perceived hazard and my real hazard. That sole option was to suspend the mission temporarily and bring our operation to the mutually acceptable, safe condition—full stop in the middle of the harbor. For him, at that time, nothing else would have worked.

Example: Never Walk under a Lift
Some of the more dangerous activities in hazardous businesses involve crane operations when lifting a load. Though the operation shown in photo 101 was conducted safely, countless accidents have happened around the world in various businesses when loads have been dropped. Yet, people continue to walk under lifts. I have always wondered why their supervisors haven't helped them to understand the technique of reducing exposure to hazards. Dropped objects can't hurt you if you never walk under them.

Photo 101: Example of a lifting operation during construction

Example: Stay Out of Blind Zones

Here's one final example that may fit within your everyday experience on the highways. Notice photo 102. You can see from the photo, taken from the passenger's side of our car, we are about to be positioned in the adjacent driver's blind zone as I pass the car with the broken side-view mirror (further increasing the size of his blind zone). To reduce exposure to this hazard when I'm driving, I avoid blind zones or minimize the time I spend transiting blind zones. When I must be in a blind zone, I am particularly attentive to the subtle signals that indicate the other driver is starting to drift into my lane, and I know what to do to react quickly and avoid an accident.

Photo 102: Approaching the driver's blind zone, made larger by a broken side-view mirror.

15. Maintain Positive Control (When Moving Objects)

During operations, equipment, people, and systems are usually in motion. Many accidents occur when something is moving incorrectly. The risk to people and equipment increases significantly when an operator—or a system, for that matter—controlling the motion of an object loses control. The operator or system doesn't have to lose control completely to cause an accident. A partial loss or degradation of control can result in an accident.

Never relinquish positive control when moving objects. Positive control implies the operator or system is making control inputs and the object is moving correctly in response to those inputs. Operators must maintain continuous control of objects through a viable method to regulate the position and velocity, which includes speed and direction, of the objects. Control must be maintained when moving objects mechanically with a motion system (forklifts, backhoes, trucks, cranes, jets, engines, pumps, or others) or manually with devices (hands, feet, poles, ropes, jacks, or others).

Through practical training in real or simulated operations, and from gaining experience in progressively more challenging environments, operators develop the skills to *Maintain Positive Control.* For the skills-based training and experiential learning to be effective, operators must first *Understand Control.*

A. Understand Control

When using a control system, operators must understand the normal modes of operation, including the automated and augmented functions used to command the motion of objects. If the system causes undesired motion, the operator must be able to override the system and compensate quickly and accurately. Additionally, operators must understand the degraded modes of operation of the control system and have an effective plan to retain positive control when failures occur.

Operators acquire the information to Understand Control from self-study and knowledge-based training provided by the organization. Understanding control is a prerequisite to Maintain Positive Control.

Example: Insufficient Control from the Digital Autopilot with LDEF Satellite Attached
For spacecraft motions requiring precise control, such as maintaining a specific attitude relative to the surface of the earth, computer performance generally outshines human performance. This is one of the reasons why attitude control of many vehicles—on the sea, in the air, in Earth orbit, and in interplanetary space—is generally handed over to the autopilot. With complete programming, computers are extremely precise in their control function. And, the computers don't get tired or hungry (as long as humans feed them electricity).

During the ascent, orbit, and reentry phases of a Space Shuttle's flight, attitude control was almost always performed by the autopilot. Just because we were not in control, the technique *Maintain Positive Control* was no less important. Because our lives were now in the programmable hands of the autopilot, I could make the case it was much more important for us to monitor the autopilot and understand the control laws the autopilot was following. We never relinquished control to any system without fully understanding how that system was performing the control function.

On STS-32, we had a unique situation in which the autopilot was not capable of safely controlling *Columbia.* The control task was so complicated the autopilot could not be programmed completely. So, the astronauts controlled the orbiter manually for three hours. Before our chests swell with unjustified pride, our need to disengage the autopilot and complete the operation manually had less to do with our individual skill and more to do with a particular design feature of the average human brain compared with a computer.

Humans are more suited for higher-level, cognitive functions using judgment, experience, and intuition. Human programming, commonly called training, can be incomplete and still be effective. For example, pilots can safely touchdown on any runway without having seen every runway. Computer programming must be complete, and on this flight, completely programming the autopilot to handle so many different configurations wasn't possible. The following was the situation.

Our primary mission was to return to Earth the Long Duration Exposure Facility (LDEF) satellite. It had been collecting data in low Earth orbit for six years. The data from LDEF would help scientists understand the orbital environment, including radiation, upper atmospheric drag, micrometeoroids, orbital debris, and other parameters that might affect the design of future satellites. For those six years, LDEF had been flying in the same orientation relative to Earth. So its forward-facing, windward side was bombarded by debris and ram atomic oxygen particles from the near-space atmosphere, and its trailing, leeward side was protected from the hypersonic impacts.

Our job was to retrieve the satellite, conduct a photographic survey of all the panels on the LDEF, as shown in photo 103, and berth it in our payload bay. To preserve the purity of the data, we protected the trailing side of LDEF from particles by using the orbiter as a shield as we sped around the earth at 17,500 mph in the upper atmosphere in the lowest fringes of space.

Preflight analysis indicated our autopilot could not be programmed to control the combined mass of the orbiter and LDEF as we changed our center of gravity and moment of inertia when moving the arm to take pictures. The LDEF, though weightless, had an inertial mass of 21,000 pounds. The autopilot couldn't handle the classic phenomenon of the rotating figure skater, in which the speed of rotation increases as the skater's arms are drawn closer to the axis of rotation. If we were

Photo 103: Using manual control with LDEF attached (Photo Credit: NASA)

rotating faster than only five thousandths of a degree per second, the autopilot couldn't control our increasing angular rate as we drew LDEF closer. The risk was losing control or structural failure of the arm and collision with the satellite. As it is with so many spaceflight scenarios, this one had the same possible conclusion: "and then you die."

So, the control task was left to the humans because we didn't want to die. With a little bit of training, we were able to control the combined masses manually instead of using the autopilot. We did what humans do best. We used judgment and intuition, fueled by training and experience, and a lot of communicating among crewmembers to coordinate the operations without losing control. The autopilot's computers couldn't do that. There were too many possible scenarios for the computer to compute. We only needed a little bit of training.

We were successful because we had the help of brilliant engineers and instructors on the ground who gave us the systems knowledge and skill before launch to use the technique Maintain Positive Control and the subtechnique Understand Control for our attitude control system.

Example: Air France Flight 447

On May 31, 2009, 228 people boarded Air France Flight 447 in Rio de Janeiro, Brazil, and didn't know they would never touch the ground again.

The crew and passengers were flying on an Airbus A330 that had three Pitot probes and six static sensors to gauge air pressure, which was converted to electrical signals and sent to air data reference modules to calculate airspeed. The flight management and guidance system, and the fly-by-wire control system, used the calculated airspeed to decide what commands to send to the hydraulic actuators to keep the airplane under control and flying safely.

Problems began over the Atlantic Ocean when the meteorological conditions created ice crystals in the Pitot probes. The icing resulted in erroneous airspeed calculations, which caused the autopilot system to disengage automatically and the flight control system to reconfigure to an alternate control law. The final accident report (ref. Ministère de l'Écologie) details what happened next.

In response to the disengagement of the autopilot, the pilots made inappropriate control inputs, relative to the actual flight conditions at cruising altitude, by pulling back on the hand controller. These inappropriate control inputs caused the airplane to stall, with insufficient airspeed over the wings. The airplane began to descend in the darkness over the ocean. For the next four minutes, the pilots continued to pull back on the control stick, rather than push forward and trade altitude for airspeed. The crew failed to make the mental connection between the incorrect instrument indications and the appropriate procedure to follow. According to the accident report, the crew never diagnosed the stall situation and continued to attempt pulling up until they tragically impacted the Atlantic Ocean, not knowing why their airplane wasn't responding to their control inputs.

The computers did what they were programmed to do, when the computers received inconsistent and erroneous airspeed indications and defaulted to alternate laws. What the accident report doesn't say is, the pilots also did what they were programmed to do. The following is why.

When stalls occur, which are exceedingly rare, airplanes are usually configured for landing, with gear and flaps down, at low altitude and low airspeed. The correct response for pilots in a stall at low altitude is to apply power to the engines, level the wings, and command the airplane to fly at optimum angle of attack, which yields a maximum ratio of lift to drag to minimize the loss of altitude. Flying optimum angle of attack is critical during recovery and is challenging for pilots.

Making the proper control inputs to recover from a stall requires skill, developed through practice in the simulator. Airline managers don't want pilots to practice recovery from dangerous stalls in the real airplane at low altitude with little margin for error. The skill of flying a precise recovery is volatile, and the pilots' ability will degrade over time without practice. Developing the proper skill to make the correct control inputs *consistently* over time requires *proficiency*, that is, high-level competence, derived from training and practice routinely and recently in the simulator.

Developing deeply proficient skills is what we did decades ago when I learned to fly, before we had fly-by-wire flight control computers. These days, training is different. With only a few lines of software code, the computer can quickly and accurately command the flight control surfaces to generate the optimum angle of attack for stall recovery. The pilot is merely required to pull back on the control stick to the aft limit, indicating the command to achieve and maintain optimum angle of attack, and the computer does the rest quickly, consistently, and accurately. Most of the time.

With the advent of computerized cockpits, training was reduced. Pilots no longer needed to spend time and expense to develop skill and maintain proficiency to control the airplane precisely. More flying time was turned over to the autopilot. Even when the autopilot was disengaged and the pilots were flying manually, the computer was still receiving the pilot's inputs from the control stick and sending computer-generated commands to the actuators, which controlled the airplane more precisely than the pilots could. Most of the time.

Reduced training was especially attractive in European organizations that provided *ab initio* training to pilot candidates, who walked in "off the street," with no previous flight training or experience. So, airlines began to train their pilots to respond to stall warning indications simply by pulling back on the stick to the aft limit. Essentially, the pilots were trained not to think about precise inputs but, rather, to pull back and let the computer do the hard work of deciding how much to pull on the aerodynamic surfaces. The computer performed the task of recovering from the stall better than the humans. Most of the time.

Yet therein lies the problem. Most of the time, the computer does better than the human—but not always. On rare occasions, the computer may receive faulty inputs, such as ice in the Pitot tubes, or be programmed with design flaws or may fail from a manufacturing defect subjected to vibrations.

This finally brings us back to the technique Maintain Positive Control (When Moving Objects), and its subtechnique Understand Control. *Positive control* means the system is responding to the operator's inputs, as intended by the operator. To make the correct inputs necessary to achieve this control, the operator must:

- Understand the control system's normal and degraded modes of operation, both of which may include automated and augmented functions;
- Know how to identify when the control system switches to different operating modes;
- Know what control inputs are appropriate in the different modes.

The organization is responsible for training the operators and giving them the correct knowledge, skills, and attitudes to apply the techniques Maintain Positive Control and Understand Control—for *all* the different modes and functions of the control system. If the operator is trained only to respond to the normal modes of operation, positive control can be lost in the rare cases when the system defaults to a degraded mode of operation. Without understanding the degraded, automated, and augmented modes of the control system, the operator won't know what control inputs to make to maintain positive control.

This is what happened tragically on Air France Flight 447. With their flight control system operating in a degraded mode, the pilots didn't make the correct control inputs to maintain positive control. After reading the accident report, we are left wondering why the pilots failed to make the correct control inputs. The answer, which is not found in the accident report, is the sum of all the pilots' training and experience led them to believe the *appropriate* input was to pull back on the stick, so they did that until they impacted the ocean.

The subtle but important distinction I'm making is this. The failure was *not* the pilots' inability to identify the correct action. The failure was all their training and previous experience told the pilots they *had identified the correct action*, which they believed was to pull back on the stick, based on their perceived environment. They simply didn't understand why pulling back wasn't working because they didn't understand the degradation in their control system and that pulling back couldn't give them positive control in their flight environment.

This example is intended to illustrate the importance of the technique Maintain Positive Control and its subtechnique Understand Control, including degraded modes of operation. These techniques are becoming more valuable in our dangerous world with increasingly prevalent automation and complex computerized control.

I believe many accident reports contain a similar common oversight in identifying failures. As Sidney Dekker has written in *The Field Guide To Human Error Investigations*, investigators tend to look for what went wrong or who made the mistake or why they failed to do the right thing. A better way to investigate, if the organization wants to prevent similar occurrences in the future, is to ask, what went right? That is, why did all the decisions and all the actions make sense to the participants at the time before the accident? Only from understanding why the operators thought they were doing the right thing can managers know how to change the conditions in the future to help operators avoid similar accidents.

Behavioral psychologists know human behavior is based on the perceived current environment and the sum of all previous experiences. Before accidents, the victims don't intentionally do the wrong thing. They assess their current situation, and they take actions based on what has worked before in similar operational situations or in training simulations. The organization is responsible for training the operators to understand control (knowledge-based training) and for giving the operators the proper experiences that will lead them to taking the appropriate actions (skills-based training) as they maintain positive control, including when the system is operating in a degraded mode of control.

B. Maintain Connection

Positive control requires a physical or mechanical connection between the object being moved and the system being used to cause motion. This connection provides sufficient ability to stop the motion or change direction to prevent undesired collisions with other objects. Simply stated, the object must be *secured* to the control system before attempting to move the object. Do not push, pull, or drop anything without the ability to control the motion.

Intentionally releasing an object (dropping to a lower level) relinquishes control and should be avoided. Pushing an object with a vehicle without the ability to pull does not afford complete control. Using a vehicle to pull an object with a rope does not afford control if a sudden stop is necessary.

Example: VSM on Forklift

Vertical support members (VSMs) are long metal columns that support pipelines above the ground. The VSM is analogous to a utility pole, which supports power lines commonly seen on roadsides.

A contractor was using a forklift to offload individual VSMs from the bed of a tractor-trailer. On the end of each VSM was a square-shaped flange, which was used to attach horizontal support structures to the VSM. Each flange had four sharp corners and edges.

To offload the VSMs, the contractor approached the side of the tractor and carefully maneuvered the tines of the forklift under a single VSM. With the longitudinal center of gravity of the VSM between the tines of the fork, he delicately tilted the fork upward so the VSM rolled gently toward the fork's elbow and nestled in the 90° angle of the fork. With the VSM in this stable position, the contractor backed the forklift away from the tractor-trailer to stage each VSM for use later by the construction crews.

One particular load of VSMs was positioned closer to the cab of the tractor in a slightly different configuration on the trailer than previous loads. To his credit, the contractor recognized this difference and assessed his task now had an additional risk. (Based on his experience, he was intuitively performing a technique I will present later: *Recognize Divergence*.)

Since the VSM might roll backward asymmetrically as he tilted the fork upward, he knew there was a higher likelihood the flange would impact the cab windows. Organizations need workers who are thinking ahead, assessing risk, and controlling hazards. Unfortunately, this contractor chose a method to decrease the risk that actually introduced a different hazard he hadn't anticipated.

To control the hazard of an asymmetrically rolling VSM that might damage the cab's window, the contractor decided to carefully position the fork under the VSM, raise the fork slightly *without* tilting, and slowly back the forklift away from the tractor with the VSM balanced delicately.

His intention was to prevent the VSM from rolling by using a feature in the control system of the forklift, which could automatically maintain the fork in a level position, with respect to gravity, even when the wheels of the forklift were rolling on uneven terrain. Unfortunately, the augmented control was not good enough to keep the VSM balanced. Just after the tips of the fork cleared the edge of the trailer, the

VSM rolled *forward* off the fork and dropped to the ground. The sharp corners of the square flange hit and ruptured a metal fuel tank under the cab of the tractor, causing fuel to spill.

The forklift operator was using some good techniques as he assessed and conducted his job. He was maintaining an awareness of risk. He astutely recognized a difference in this particular load. He controlled one new hazard but overlooked a second. The accident might have been averted had he applied the subtechniques Understand Control (including augmented functions) and Maintain Connection before he backed away with the VSM balanced precariously on the fork and not properly secured (connected).

Example: Dropping a Chain Bucket

A supervisor and his two-person crew were conducting a job to install a chain hoist in the overhead (roof) of a hangar. The work was going well. Using the approved procedures with a valid work permit, supported by risk analysis, the crew had successfully completed most of the complicated and heavy work to install the hoist. Near the end of the shift, only one task remained. To complete the job, the crew needed to raise and install in the overhead a chain bucket intended to contain excess lengths of the chain when the hoist was used during future operations.

Before conducting this final task, the supervisor and his crew reassessed the hazards they would face. The task involved multiple steps. The workers had to raise themselves and their tools to the ceiling in the basket of a lifting device known as a man-lift. A similar device is shown in photo 104. The chain bucket, which weighed 190 pounds, was to be raised to the ceiling separately and positioned for installation by the workers.

The supervisor decided if they modified the procedures specified by the approved permit, they could reduce risk. Rather than conducting separate lifts, each with hazards, the supervisor decided they could complete the task more safely in only one

Photo 104: Man-lift device used to hoist people and equipment during construction operations

lift by bringing their tools and the chain bucket with them simultaneously in the man-lift basket to the overhead.

Before attempting to conduct this final task with modified steps, the supervisor and his two-person crew discussed how they would manually load the chain bucket into the basket. The chain bucket was an unwieldy piece of equipment, which had to be lifted over the rails of the basket, which was about four feet deep. All three crewmembers would lift the bucket by hand over the rails. The supervisor would give a verbal command, and the three people would release the bucket simultaneously, dropping it into the basket.

That was the modified plan. This is what happened when they executed it. As they dropped the bucket into the basket, a metal fitting welded to the side of the bucket impacted the supervisor's hand against the railing of the basket causing an injury that required medical attention. The supervisor then had to admit he and his crew were conducting the task with a process not in accordance with the approved work permit.

Managers in the organization were disappointed with the supervisor who failed to exhibit proper leadership behaviors by directing his two-person crew to conduct work in violation of the approved permit. Had the supervisor not been injured and the unauthorized work continued, the managers realized the crew might have violated the six-hundred-pound maximum load limit in the man-lift basket, which would have contained two crewmembers, tools, personal protective equipment, and the 190-pound chain bucket.

The supervisor acted in violation of organizational policy. I believe the managers were justified in their disappointment, especially considering the higher responsibility of leadership bestowed on the supervisor. But, I also believe many organizations need to do more than merely enforce conformance with policies. Middle-level managers and front-line supervisors should be teaching and influencing good work habits and proper techniques for operating excellence in the field. When the supervisor briefed something like, "On the count of three, we will drop the bucket," loud, internal warning sirens should have been activated automatically in the heads of the supervisor and his workers. Dropping equipment intentionally is never a good idea, especially hardware that weighs 190 pounds, from a height of four feet. In my operator's mind, that was the greater deviation from operating excellence.

Managers in many organizations will continue to direct efforts to enforce conformance with rules-based policies and procedures. Some accidents will be prevented. Managers in the best organizations will supplement their efforts on policy conformance with efforts to teach good work habits through principles-based *Techniques for Operating Excellence*. More accidents will be prevented. When principles-based techniques are mastered, all accidents will be prevented.

C. Prevent Unintended Collisions

1. Maintain Accurate Knowledge of the State Vector of the Object
The operator must always maintain accurate knowledge of the state vector, that is, position and velocity, of the object being controlled. The state vector is changed by the outputs of the control system, including uncommanded outputs and undesired motion caused by failures in the system. Failing to monitor the outputs of the control system and the motion of the object may result in unintended collisions. (This subtechnique specifically relates to preventing *unintended* collisions. In operations, some collisions are intended, such as landing, docking, grappling satellites, mounting structures, connecting equipment, capturing targets, connecting tractors with trailers, and others.)

Example: Construction Cranes over Paris Skyline
Using a construction crane effectively, efficiently, and safely can be challenging for a crane operator. Structural loads on the crane can vary greatly as payloads are raised and lowered. The internal dynamics of the system, including flexibility of the structure and longitudinal elasticity or lateral motion of the cable, can excite instabilities or fluctuations in motion of the payload. External environmental factors, such as steady or gusting wind and turbulence, can adversely affect the motion of the payload. The center of gravity can shift widely as the crane is rotated horizontally and vertically or if the payload begins to oscillate. Interface issues can arise during construction operations with multiple cranes and overlapping

Photo 105: Construction cranes over Paris skyline, 2011 (Photo Credit: Kelly D. Wetherbee)

envelopes. The different control laws in the crane's control system may not be able to accommodate all these variations. The resulting outputs of the control system may seem unpredictable or unexpected.

The crane operator must understand all these factors all the time during operations. When the crane is operating in a construction site in a populated area, as depicted in photo 105, the lives of the crane operator and the spotters are not the only ones at risk. If the crane suffers structural collapse during operations in a city, many people can be killed. Maintaining an accurate understanding of the state vector of the crane equipment and the payload, including predictions of their future position and velocity, can save lives.

Example: Construction Cranes over Earth's Skyline
On STS-113, I had the privilege of operating the *Endeavour's* crane (the Remote Manipulator System, or arm) to lift the P-1 truss element from the payload bay of the orbiter and hand off the element to astronaut Peggy Whitson using the crane on the International Space Station during construction. Photo 106 shows her crane just to the right of center in the picture.

Though the P-1 truss element was weightless, it had an inertial mass of 27,000 pounds. With only 1.5 inches of clearance on either side of our fifteen-foot-wide payload bay, I had to be diligent and vigilant as I slowly and carefully lifted the truss element using the crane. Had I made a control error or if the crane experienced a control-mode failure and the truss impacted the structure of the payload bay, the 27,000 pounds of mass would have torn right through the orbiter's structure like a massive eighteen-wheeler going through a garage wall on Earth if its brakes failed.

Photo 106: Construction cranes over Earth's skyline, 2002
(Photo Credit: NASA video screen capture)

During the entire operation I expected such a failure. With the benefit of the exceptional training from skilled instructors before launch, I was prepared to take immediate corrective actions to prevent disaster when using the crane in space. No failures occurred and the operation was successful.

2. Maintain Accurate Knowledge of Local Environment

The operator must understand the environment around the object being moved by the control system. Failing to understand what other objects and structures are near the object being moved may result in unintended collisions. When the driver of a vehicle backs up without being able to see the environment behind the vehicle accurately and completely, an unintended collision may occur. Outside observers or spotters are used when drivers cannot see the local environment around a large moving vehicle.

Example: Wing walkers

In the aviation industry, wing walkers, or spotters, are employed to help the pilots avoid collisions on the ground when objects in the local environment are not easily seen from the cockpit. The job of a wing walker may seem simple enough; not much ever goes wrong. But occasionally, dangerous situations occur. Constant attention is required. Will the wing walker have sufficient awareness of the environment and a heightened state of mental readiness to take immediate corrective action to prevent an accident on the rare day when an obstacle suddenly presents a conflict?

You might assume since aircraft are towed at low speeds the likelihood of collisions would be reduced. Still, unintended collisions occur. Countless aircraft, including the Space Shuttle, have been towed or taxied at a speed of one mile per hour into objects. In 2005, *Endeavour* was being moved into position in the Orbiter Processing Facility when it contacted a scaffold, which was in a non-standard position. Although a radio communication problem contributed to the incident, the unintended collision may have been avoided if the move personnel had maintained accurate knowledge of the local environment.

3. Anticipate Potential Conflicts or Collisions

When controlling an automobile or truck in an environment with other moving vehicles or objects, the driver must be able to predict the motion of all the other objects. The skill required to integrate the entire scene or environment and avoid potential conflicts or collisions must be practiced. Operators who control moving objects must develop and demonstrate capability through a qualification program. After being certified by an assessor in the qualification program, the operator must maintain proficiency (deep competence, with recent experience) through a recurrent program of practice or operations to ensure the volatile skills to *Anticipate Potential Conflicts or Collisions* are not lost.

4. Assume the Worst

To avoid collisions or accidents when moving objects, always *Assume the Worst* will happen. When driving a car on a curved road, assume an obstacle is just out of sight around the corner. *Do not drive faster than you can see (and respond).* In other words, preserve the ability to avoid obstacles that suddenly appear in your path of motion. Assume other drivers will run red lights or stop signs. Drive in such a way to avoid the violators. Assume your control system will fail at the worst time, anticipate the consequences of the failure, and use a plan of operations that preserves options to maintain control of the object. When using any system with augmentation or automated modes of operation, anticipate what will happen when the augmentation or automated system defaults to a degraded mode of operation. Have a plan to reduce the consequences of these failures if they occur at the worst possible time during operations.

Example: 45 mph on the Curve

Notice the scene depicted in photo 107. I almost never drive at the speed limit on winding mountain roads. The problem isn't so much my skill. The problem is, at forty-five mph, I can't see far enough around the corner. So, my personal speed limit is defined by never driving faster than I can see. I also never drive faster than I can trust. I want to be able to see and respond to obstacles, hazards, or tired drivers who are veering into my lane. What if the approaching truck blows a tire or spits a load just before I arrive? If I drive into an accident, I shouldn't blame the tire, the load, or the state official who decided forty-five mph was a safe speed limit.

Photo 107: Never drive faster than you can see (and respond).

Each of these fifteen techniques has the potential to prevent accidents and improve performance but only if executed with an appropriate mental attitude that reflects operating excellence. The techniques work well when used singly. They work better when used together with other related techniques. They work best when the operator exhibits the desire to master the techniques to stay alive and help the organization achieve greater success.

Now I'll turn to the second fifteen techniques that can help operators on the social side of the sociotechnical system used to control risk in dangerous endeavors.

Techniques for Operating Excellence 16–30

16. Balance Confidence with Humility (Individual and Organizational)
17. Communicate Effectively and Verify Communications
18. Be Prepared Mentally
19. Be Mindful During Operations
20. Think Fast and Act Deliberately
21. Recognize Divergence
22. Share and Challenge Mental Models
23. Challenge "Go" Deliberations
24. Be Assertive (to Authority) When Necessary
25. Be Cognizant of Limitations (in the Sociotechnical System)
26. Assess Competence (in Team Members)
27. Acknowledge (Personal) Weaknesses
28. Admit Errors
29. Use Methods to Aid Weak Prospective Memory
30. Demand Operating Excellence from Myself First (Then Inspire Others)

Chapter 5

PERSONAL TECHNIQUES FOR OPERATING EXCELLENCE
Part 2: The Social/People Side

The *Techniques for Operating Excellence* on the people side of the sociotechnical system are more challenging to define and explain than those on the technical side. I suppose this is expected. People are softer and more challenging to define than the sharp-edged technical processes. So it is with these techniques on the people side. These fifteen people-side techniques are softer in definition and contain more overlaps than those on the technical side. Computers are rational and behave consistently. (So far, anyway.) People can be irrational and inconsistent. Emotions muddy the behavioral waters in people. But, that's the beauty of being human. We don't want to behave like computers. Emotions breathe life into our actions. Emotions create the individual spark in each of us that ignites creativity and productivity. Emotions, when mastered, help us achieve greatness and accomplish much more in our dangerous world.

And that is the beauty of these fifteen techniques on the people side. These techniques will be more challenging to discuss, teach, practice, and master. But, it will be worth the effort. If mastered, these techniques will motivate people to work together and will inspire operating excellence, yielding better performance to achieve higher-quality results. These social-side techniques will enable personnel at every level in your organization to prevent accidents, save lives, protect investments, generate higher-quality results, and ultimately take your company higher as you continue to deliver value to society for the long term.

Techniques for Operating Excellence 16 through 30 are intended to help operators control risk when conducting dangerous operations and are specifically related to enhancing interpersonal relationships and operating with people organized in groups to achieve missions.

16. Balance Confidence with Humility (Individual and Organizational)

In today's dangerous world, confidence is needed to enter a profession in pursuit of missions in which the consequence of failure can be death. Humility creates successful operators and managers by influencing the right choices, decisions, behaviors, and actions. Confidence gets you into the arena. Humility keeps you alive long enough to win the game.

To operate effectively in hazardous environments, operators and managers must be highly confident, capable of withstanding psychological and physical stress while still making the right decisions and taking the right actions. Confidence is a reward that comes from training and preparation. But to enable the highest performance consistently and avoid accidents, an operator must *Balance Confidence with Humility*. Humility allows the individual to capture errors, learn more, and stay alive. Supreme confidence must be tempered with healthy self-doubt. Without confidence, operators will make mistakes. Without humility, operators will not realize they are making mistakes.

Display humility. Mitigate intimidation. Cultivate relationships. Better decisions and higher performance will result. Embrace the healthy self-doubt that derives from humility. This creates an awareness of vulnerability and the ability to capture errors, stay alive, and improve performance.

Before critical operations, good operators channel their self-doubt in a productive way by asking themselves, "What have we missed?" or, "What have we failed to do?" or, "What mistakes have we made which can be corrected before it's too late?" Operators and managers should always question the processes they are following to achieve the mission. This questioning may provide the only opportunity to identify issues with the processes.

After the *Columbia* accident, I was responsible for helping the managers, engineers, and operators at the Johnson Space Center make changes to create and sustain a culture of operating excellence. During one meeting, I was discussing with some managers the appropriate behaviors to display during contentious meetings involving difficult technical issues. One of the senior managers asked me, "Are you saying we have to be nice to people all the time?" In the moment my answer was,

"No, but we need to treat everyone with respect at all times." Later that day, I realized my answer should have been, "You need to behave in the best way to influence better decision-making and encourage operating excellence."

If the best way in a particular situation is to be nice, then be nice. In decisions involving dangerous activities, human nature likely will impel the bosses occasionally to issue orders or make directive statements that don't appear nice or agreeable. Those bosses must do what they feel is necessary at the time to make the best decisions to create and maintain operating excellence. But after the meeting or operation, the leader would be wise to return to the individuals in the team and exercise good leadership by verifying any overly authoritative or aggressive behaviors didn't damage the climate of good decision-making and operating excellence. The best leaders don't simply make the best decisions in the moment. The best leaders use appropriate behaviors in the moment to foster the best decision-making in the future.

Operators or managers who display confidence without humility are found in many organizations. They will be easy to identify. This attribute of confidence without humility appears as overconfidence. Find these people and avoid being trapped in their trailing wake of dysfunction and cultural damage.

Never blame others. Have the humility to look inward and understand how your own decisions, actions, inactions, and failures have contributed or will contribute to a problem.

True listening occurs through humility. Have a desire to understand others' opinions. Some people are masters at displaying body language that indicates they are listening, but when it is their turn to speak, it is obvious they were simply rearming or reloading for their own verbal salvo. For better decisions, value diversity of thought. Encourage questioning. This comes from a place of genuine humility. Good managers always consider the possibility they have the wrong opinions or interpretations.

Finally, remember this technique applies to all individuals in the organization—and to the collective organization. Leaders at every level can benefit from mastering these Techniques for Operating Excellence. This one in particular, *Balance Confidence with Humility*, is especially important for the top executives. If they set the proper tone by demonstrating individual humility, they can create organizational humility. Executives and managers throughout the organization will recognize they don't have all the answers. In a demonstration of humility, these leaders will encourage a continuous search for vulnerabilities by all personnel. Operating excellence comes from confidence balanced with genuine humility in every leader. When the organization collectively masters this technique, personnel will be able to improve safety, short-term operating performance, medium-term productivity, and long-term shareholder value.

Example: STS-52 Launch Hold—Confidence and Self-doubt Paradox Reconciled
For the first fourteen years in my career as a naval aviator, I operated successfully with an apparent contradiction in my mind. In one part of my brain, for as long as I could remember, I felt supremely confident in my ability to control hardware. I'm not sure where this came from, but I always felt I could control anything that floated, glided, rolled, or flew—no matter how high or how fast. But, as powerful as that feeling was, I concluded my confidence was an unjustified and unearned emotional feature in my mind.

After each jet flight early in my career, I didn't think I performed well enough. I felt I *could* be a good pilot, but I didn't think I *was* a good pilot. The rational side of my brain was leaving me with self-doubt. In the training command, during my first night formation flight, I thought I was so shaky on the controls I convinced myself I would receive a failing grade. After the flight, I sat down for the debriefing and prepared myself for the worst from the instructor. His only comment, which he duplicated on the training form, was, "Smooth flight; great hands on the controls."

That only served to increase my confusion. Though my confidence grew, so did my self-doubt. Maybe I really was better than other pilots. But that didn't matter if I wasn't good enough to stay alive in dangerous environments.

Throughout my training and later in the fleet, I continued to receive high grades and positive evaluations. And I continued to think the generous assessments were unjustified. Over time, I learned to live with the strange dissonance in my mind. I began to realize the dilemma of simultaneous feelings of confidence and thoughts of self-doubt was actually an asset in my job. My confidence helped me fly without worry, and my self-doubt caused me to study and prepare more for each flight.

On the emotional side of my mind, I always felt comfortable when the missions got dangerous. On the rational, doubting side of my mind, I knew I didn't know everything I needed to know to perform well. This motivated me to study harder and to prepare more extensively before each flight or mission. Later in my career, I was fortunate to have a wife who tolerated every weekend I spent deeply buried in aircraft or Space Shuttle training manuals trying to get smarter.

Yet, for years I wondered, *Which is it?* Am I the best pilot, or am I the worst pretender? Which is more useful: confidence, even if it's a misplaced feeling, or self-doubt, even if it's justified thinking? I found out on November 1, 1992.

I was on the launch pad, strapped in for my second flight on *Columbia* but my first time from the left seat, as the commander of STS-52. My crew and our launch team had completed most of the prelaunch sequence, and we were holding at a countdown time of T minus nine minutes. The launch director told us we would be holding for an extended period because of high crosswinds at the Kennedy Space Center runway, which we would need to use only if we suffered a main engine failure during ascent. The gusting wind was a violation of one of the flight rules specified in our Launch Commit Criteria. Because the launch window was more than an hour long, we could afford to settle in and see if the winds died down before we needed to scrub the mission for the day.

On this launch morning, as the crew of STS-52 waited, the crosswinds were gusting to twenty-two knots, seven knots over the recently approved limit. I was the weather coordinator at the Cape for seven launches, and I knew the crosswinds likely would not decrease to less than fifteen knots this day. After listening for a few minutes to the discussions by the launch team on the radio, I told my crew on our intercom, "There is no way we are going to launch unless the winds decrease."

With the checklist completed, nothing to do, and nowhere to go, I fell asleep in my seat. This was not a good thing to do. I startled myself awake when my left elbow jerked and inadvertently hit a circuit breaker panel. So, I decided I had better keep my mind active and stay awake. What I thought about were the following three things.

First, we were incredibly privileged to be sitting on top of a rocket, built by hundreds of thousands of truly dedicated people around the country who put their hearts and souls into America's space program.

Second, I completely trusted those highly dedicated people, and I believed they had given us a space vehicle that would operate correctly. And, those same people trusted me to fly the mission and bring *their* vehicle back to them in one piece. I had better honor the trust they placed in me by not failing.

My third thought brings me to the point of this example for this technique. After all those years in cognitive conflict, as I sat there on the launch pad, on top of a rocket, on top of the world, I finally was able to reconcile the apparent contradiction in my mind about which was misplaced, my confidence or my self-doubt, and which one I needed more.

The answer came to me, as I thought about all those people supporting us and I realized I did not want to let them down. In a dangerous business, you need both, supreme confidence on the emotional side *and* healthy self-doubt on the rational side. These are not in conflict; they are synergistic. I had every right to be supremely confident. The crew of STS-52 had the finest training from outstanding instructors using incredible facilities. At that moment, we were the most highly trained crew on the planet. We were ready. But, I also knew I could make a mistake in any individual operation, so I had better preserve a healthy amount of self-doubt. That humility, to know I could make a mistake, is what I will need to prevent mistakes and keep my crew and me alive.

I finally realized the value of balancing confidence with humility. The operational connection between the two can be illustrated with this self-assessment question: I know I am capable of operating successfully today, but am I performing successfully in this moment?

This distinction became clear as I thought about that dark night behind the ship years earlier when I landed hard and realized some hazards were in my head, not in the external environment. As a naval aviator, I had the confidence to execute arrested landings aboard a ship under any environmental condition. But on that particular night I failed to execute an acceptable landing when I had forgotten the value of humility and I neglected to control my mental processes.

As I waited on the launch pad, these thoughts coalesced. Confidence allowed me to climb through the hatch and strap in, knowing we were ready. The humility to know I could make a mistake will cause me to pause just long enough to verify the accuracy of my decisions and the appropriateness of my actions. Together, supreme confidence and humility will allow me to think clearly after liftoff as my brain is accelerated to the back of my head under the power and fury that is a rocket launch and still make the right decisions if things begin to go awry.

Balancing confidence with humility is how I will keep my crew alive, save the vehicle, and complete the mission—in that order.

In case you are wondering, I was right about the weather conditions. The winds remained high, with a peak gust of twenty-two knots, throughout the entire launch period.

And, I was wrong about the launch decision. They launched us anyway. Photo 108, also used on the cover of this book, shows the spectacular results of the launch decision. But, I'll save the story of how we came to be launched on STS-52, with winds exceeding the crosswind limit, to illustrate another of the people-side Techniques for Operating Excellence—*Be Assertive to Authority when Necessary.*

Photo 108: STS-52 ascent—balancing confidence with healthy self-doubt while riding 7.7 million pounds of thrust (Photo Credit: NASA)

Example: Italian Flight Controllers
During more than five thousand hours of flight time in airplanes, I have heard many pilots arguing with flight controllers over the radio about various requests from the pilots. In most cases, the controllers delayed or denied the pilots' requests. The more demanding the pilots became, the more reluctant the

controllers seemed to be to grant requests. To me, the concept seemed simple. In my entire career, I have never argued with a controller. And, in my entire career, I have never had a request denied.

Here's an example that always makes me smile. Photo 109 is a picture I took from the cockpit of my A-7 prior to ingressing on a low-level training flight from the Mediterranean Sea. Several times during our seven-month cruise, the A-6 and A-7 attack squadrons from Air Wing One sent multiple airplanes on training exercises over friendly territory to simulate a coordinated ground attack mission.

Photo 109: Low-level ingress route on a training mission from the USS *John F Kennedy* in the Mediterranean Sea

My favorite controllers were the Italians. One morning during a large training exercise, I approached Italy from the sea. I could hear bombardier-navigators from the A-6 airplanes arguing with the Italian controllers, pleading their cases and attempting to acquire clearance to overfly the country. The more the navigators argued about getting to their targets on time, the longer they and their pilots were stuck in extended holding patterns. The Italian controllers didn't seem to understand the radio transmissions from the navigators and definitely were not responding to the urgency of their requests.

The following represents a typical exchange, from thirty-five years ago, between the Italian controllers and me, using the radio call sign Decoy 407:

Decoy 407 (in a cheery voice): "Catania Approach, Decoy 407; Good morning!"
Catania Approach (in a thick Italian accent): "Ahh, Decoy 407! Nice to hear you again! And how are you today?"
Decoy 407: "Fine! How are you?"
Catania Approach: "Wonderful! You know where you are?"

Decoy 407: "Yes."
Catania Approach: "You know where you want to go?"
Decoy 407: "Yes."
Catania Approach: "Roger—you cleared!"

(Actually, if I was a little more confident, I might have slipped in a quick, "*Buongiorno!*" But I was still a new guy and didn't know if a violation of the English-only rule would be overlooked in the interest of maintaining good international relations.)

Part of me felt sorry for the A-6 pilots and navigators as I flew past them stuck in their endless holding patterns. They never did figure it out. Balance confidence with humility. Keep your confidence inside and let your humility shine outside to highlight others. Give people a smile, and they will be only too happy to give you something in return.

17. Communicate Effectively and Verify Communications

A. Communication During Operations

Doing work separately and simultaneously, teams of people conduct complicated operations in hazardous endeavors. Accurate transfer of information is required to successfully conduct and coordinate these activities. This information transfer is best accomplished by communicating, mostly orally, but sometimes with visual signals, symbols, or writing. The communication is effective when there is accurate and timely transmission of information and, more importantly, accurate receipt, interpretation, and understanding of the information.

Effective operations cannot be accomplished without effective communications. The penalty of errors will be degraded performance or failure to complete the mission. Miscommunication can also cause accidents that result in damage to equipment and injuries or death to people.

Communication must be accurate, whether the teams are reviewing past activities, coordinating current activities, or planning future activities. Of these three, the requirements for communication are more stringent when operators are coordinating current activities. Because of the constraints on the amount of time available in current activities, in addition to being accurate, the communication must be efficient.

Throughout my aviation career, on nearly every airplane flight, I heard at least one instance of confusing or misleading communication involving nonstandard or incorrect terminology between another pilot and flight controllers on the radio. Research has shown some type of crew error is made on an average of four out of five airline flights. Often that error involves miscommunication between pilots and controllers. Sometimes, poor communication results in fatal accidents.

Good pilots who have any doubt about what the controller said will verify the information. Even when there is no doubt, operators employ the good practice of using terminology that clearly eliminates potential doubt or confusion in the other person's mind. Reading back pertinent information between the pilot and the controller is a method of verifying communication.

Example: Effective Communication in the Oil Field

Effective and accurate communication is critical in any hazardous business. To illustrate the point, a quick example from the oil fields is relevant. A board operator in the control center might call on the radio to a field operator out in the plant and give a verbal command to open a valve. The board operator should never say, "Open the valve a crack." This command is fraught with ambiguity and uncertainty. To whom is the command directed? Which valve is desired? How much is "a crack"? Which direction is open? Rather, the board operator should use better comm technique by saying, "ULC 2 ops, board ops; Mike, open valve 27 one quarter turn counterclockwise."

Mike, who now knows the communication is intended for him, should not say simply, "Roger." What if the radio transmission was intermittent at an unfortunate moment and Mike only heard "valve 20 . . ." instead of "valve 27"? To eliminate all possibility of error, Mike should use a better technique by responding, "Roger, ULC 2, valve 27, one quarter counterclockwise." With that efficient response, both operators know the correct information was transferred and understood.

Example: CAPCOMs

For operational reasons, effective communications between crewmembers in space and flight controllers in the Mission Control Center are vitally important. In the extreme, the success of the mission and lives of the crew can depend upon critical information conveyed accurately between astronauts and controllers. For psychological reasons, crewmembers often appreciate having a pleasant conversation with team members on the ground that provides an emotional link from the isolation of space to humanity back home on Earth.

When traveling at five miles per second, in addition to being effective, the communication must be efficient. Over the years, flight crewmembers have developed a lexicon and a way of communicating over the radio that conveys a high ratio of information content to words spoken. Of course, this high efficiency is predicated on complete accuracy.

Since May 5, 1961, when Alan Shepard first launched into space, crews in flight spoke only to other astronauts in the Mission Control Center, who were providing ground support from a position on the console known as capsule communicator, or CAPCOM, in space lexicon. Over the years, this tradition of pilots speaking to pilots was valuable and successful. Astronauts on the ground, who had trained in roles as backup flight crewmembers, were intimately familiar with the flight operations the astronauts in space were conducting. Through shared experiences, astronauts on the ground could infer the operational and psychological state of mind of their office mates in space.

These days, with continuous presence in Earth orbit and generally more relaxed operations on the International Space Station, the position of CAPCOM has been opened to highly skilled and operationally savvy engineers who are not astronauts. This practice has been equally successful.

Photo 110 shows two of the CAPCOMs who supported us on mission STS-113, astronauts Barbara Morgan and Mike Massimino. Barbara is a schoolteacher, who taught in Montana and Idaho, and was originally selected in the Teacher in Space Project. She was the backup flight crewmember to Payload Specialist Christa McAuliffe, who died tragically in the *Challenger* accident. Twelve years later, Barbara applied to the astronaut program and was selected by NASA as a mission specialist in the seventeenth group of astronauts. Barbara was a superb astronaut with an outstanding work ethic who later flew on STS-118. As director, Flight Crew Operations, I tried to encourage other astronauts in the office to emulate her professional demeanor and drive for operating excellence.

Mike Massimino was an engineer from New York who flew on two greatly successful servicing missions to refurbish the Hubble Space Telescope and conducted four very complicated and highly productive spacewalks. In addition to

Photo 110: Mission Specialists Barbara Morgan and Mike Massimino, CAPCOMs for STS-113 (Photo Credit: NASA)

being one of the best extra vehicular activity crewmembers, Mike was also one of the most humorous.

My crews worked hard to master the art of communicating during space missions. We had a few simple guidelines. I have already mentioned our communications must be efficient and accurate. The goal was to convey the most information in the least number of words. The words used must be clear and understandable on a radio that might contain static. Words with hard consonants are more easily understood than words with soft consonants. If there were any possibility of misunderstanding, either person would verify or retransmit the comm. For critical exchanges between the ground and space, key phrases were always repeated regardless of the original clarity. With this technique, each team member had no doubt the other team member understood the correct information.

When I was leading the branch of the Astronaut Office that included the CAPCOMs, I created one additional communications protocol for use as a courtesy to the flight crews. The CAPCOMs were allowed to respond to whatever level of humor the flight crew initiated, but were never allowed to initiate or elevate the level of humor. Without being in space, the CAPCOMs didn't know the state of mind of the flight crew. Support personnel on the ground did not want to risk subverting the authority of the commander, who might be using leadership skills to create a heightened sense of professionalism necessary during certain situations with the crew. Interrupting a serious tone on board with a joke from a tone-deaf CAPCOM trying to be funny would be poor form.

Years later, as the operational search director responsible for leading the effort to search for and recover the human remains of the *Columbia* crew, I learned the highly professional special agents from the Federal Bureau of Investigation had a nearly identical communications protocol in place. The special agents were sensitively attuned to the psychological state of mind of the grieving organization they were supporting in times of tragedy. In NASA, we had just lost seven friends. Humor is powerful and can help operators recover emotionally when operations are critical, but the special agents waited for NASA personnel to initiate and indicate the desired level of humor before they responded in kind.

––––––––––

Example: Comm Rules During Critical Operations
On the five missions I commanded, I enforced specific communication protocols. During the critical phases of flight, including ascent, rendezvous, docking, reentry, approach, and landing, I only wanted to hear operationally relevant communication from my crewmembers on the intercom. I made a personal commitment to each astronaut that I would listen to everything they said, process their input or request,

and respond to them thoughtfully. With my commitment to them, I asked only that they fulfill one additional responsibility. Because I would devote attention to their communication, I asked them to make sure their comments were relevant to our current operation. I only wanted to divert my spare brain cells to information related to the performance of our engines, the operations of our computers, the status of our life support systems, or anything else regarding operations I might be missing. I did not want to be distracted by irrelevant comments about the view or, "Wow, what a ride" or, "Congratulations, rookie, we're above fifty miles; you're officially an astronaut!" Nice sentiments, but unprofessional.

Photo 111 shows our mid-deck crewmembers strapping in before the launch of STS-86. Notice the mid-deck crewmembers had no computer displays or external windows to monitor operational status of the ascent into space. Out of operational necessity and courtesy during ascent, most flight-deck crews on other missions kept their mid-deck crewmembers informed by periodically communicating information about the ascent performance and trajectory. This kept the mid-deck crewmembers mentally engaged in the operations in the event they needed to execute critical procedures on the mid-deck during an emergency that required crew actions.

But I always thought providing dedicated communication to the mid-deck crewmembers was not the best way to operate. Though remaining mentally engaged was an operational necessity for the mid-deck crewmembers, they didn't need separate, dedicated communication, which could have distracted the flight-deck crewmembers from their own cognitive responsibilities. The mid-deck crewmembers were skilled enough to infer the status of the ascent from the operationally relevant and necessary communication among the flight-deck crewmembers. In our high-risk operation, all communication from my crewmembers was focused exclusively on high-value information.

Photo 111: Strapping in on the mid-deck, STS-86 (Photo Credit: NASA)

I had one other set of rules during the prelaunch countdown that is worth mentioning. Humor is important for people to release tension in stressful situations. But we had to balance our use of humor with the operational requirements for communicating with each other and the launch control team. So, my rules for the use of humor on the launch pad were:

- Before the scheduled hold period at T minus twenty minutes in the countdown, humor was allowed, as long as it was not offensive (some ground personnel could hear our intercom), and the humorous story was suspended immediately for any other communication being transmitted to the crew.
- Between T minus twenty minutes and the scheduled hold at T minus nine minutes, only one-liners could be spoken. No long anecdotes were allowed because operational communication was increasing in frequency and importance as we approached liftoff.
- After picking up the countdown from the nine-minute hold and all the way through liftoff and ascent into orbit, no jokes were allowed. I only wanted to hear operationally relevant comments.

Some crewmembers wondered early in training why I was so specific with rules and so particular about how we communicated. I had only one reason, and it was bigger than mere communications. I wanted to create a sense of professionalism and operating excellence in the way they processed information that would help us stay alive and accomplish dangerous missions. Professionalism and operating excellence are largely mental attitudes that come to life in operators after years of high-quality training and superior mentoring. I didn't have the time to make drastic changes in the crewmembers' existing ways of thinking. Additionally, as a commander, I never thought I could influence and improve their mental attitudes by directly asking them to display better attitudes.

So my communication rules were a quick, yet slightly indirect, way for me to focus the crewmembers' attention on the operationally relevant technical information necessary to make the best decisions in a hazardous environment. By specifying what they couldn't talk about, I was eliminating what they shouldn't think about. As our training progressed, I could tell they were thinking about the right things and were quicker and more accurate in their observations, decisions, and actions in our high-risk occupation with no margin for error. With their confidence growing, they recognized the value of eliminating distractions and staying mentally focused on operationally relevant tasks. Each of my crewmembers seemed to appreciate the professional way we communicated and performed on our missions. To some extent, this operating excellence was catalyzed by the rigorous and specific rules for our communication.

Providing rules for communication and humor, with rationale, was a way to influence the critical thinking skills and proper mental attitude necessary in our high-hazard occupation. In the end, I couldn't have been more proud of the way my crews communicated, the decision-making skills they developed, and the professionalism and operating excellence they exhibited during the critical phases of the missions.

Example: Comm Agreements with Flight Director and Flight Crew
In the previous examples, I wrote about the need to be effective and, depending upon the criticality of operations, efficient when communicating. Operators can't waste words or time when dealing with high risk. As a commander, I promulgated rules that specified when the crewmembers should listen, rather than talk, because information from the Control Center was necessary and sometimes had a higher priority. I don't, however, want to convey an impression I think communication during dangerous or critical operations should be minimized or limited only to be efficient.

The most important goal of communication is to transfer information between operators. I would rather overcommunicate than undercommunicate. My philosophy of communicating so much information may have differed from other astronauts. Here is an example to illustrate.

Photo 112: Flight Directors Wayne Hale and LeRoy Cain in the Mission Control Center, Johnson Space Center, Houston, Texas, during STS-113 launch countdown operations, November 23, 2002 (Photo Credit: NASA)

Example: Radio Communication on Final Approach

Photo 112 shows Flight Directors Wayne Hale and LeRoy Cain during the launch countdown operations on STS-113. Wayne was our lead flight director. I had worked with him before, and we shared a special bond over the radio. In space, my life and the lives of my crew depended on the decision-making abilities of the flight directors. Wayne was incredibly good, and I enjoyed flying missions he controlled.

One of the preflight agreements I made with Wayne concerned how we intended to communicate on short final just prior to landing at the end of the mission. This is a time most crews wanted to minimize communication with the ground so they could focus on the critical task of landing the orbiter because there was only one chance. Screw it up and we die.

After making the final turn on the approach to landing and acquiring the runway visually, other commanders in the Space Shuttle program traditionally called, "Runway in sight," over the radio. The commanders used this call to convey two things to the flight controllers in Mission Control. The first meaning was obvious—we have the runway in sight and will be flying visually, as opposed to flying on instruments. The second meaning was tacitly understood by the CAPCOMs and flight directors. This call also meant "please limit your transmissions to us from now on; we need to concentrate on landing, and we no longer require your advice on navigating to the runway."

I had a different philosophy. Before flight, I told Wayne Hale I would not call, "Runway in sight," even after my copilot and I had acquired the runway visually. I did not want the Mission Control team to limit their radio transmissions to my crew and me. In this case, I am not judging the motivations of other commanders or crews. If they wanted to concentrate on their own navigation and internal crew communication and didn't want to be distracted by external comm from Mission Control, that reflects their way of operating. But I valued any information that Wayne wanted to pass along to my crew and me, even if I could see the runway. I might be making a mistake on final approach, and he might say just the right thing to help me mitigate the consequences. During this time, and any other time in the mission, I wanted him to say anything he thought was important. If I became overloaded and couldn't process the communication, I would mentally filter out his transmission.

In the end, we didn't make a mistake on final approach that required Mission Control to say anything. During the entire mission, Wayne and his team did an exceptional job for us on STS-113. With their help, the flight crew and ground teams made the fewest errors of any flight I supported, as a member of the flight crew, a CAPCOM, or director, Flight Crew Operations.

The goals for communication when operating in dangerous environments are:

1. *Communicate Effectively*. When communicating, operators should endeavor to maximize the quantity and quality of information content. Operators should convey as much accurate information as possible so all members understand as much as they can about the operation. More importantly than being accurately transmitted, the communication must be accurately received and understood. All terms must be defined, used correctly, and understood commonly. Communication is ineffective, even after being transmitted perfectly, if the information is not received or understood accurately. Communicators must devote some of their communication to verifying receipt of information.

2. *Communicate Efficiently*. When operations are time constrained, under high risk in a rapidly changing environment, communication must be efficient. This is why pilots, and operators in other hazardous occupations, have developed a common lexicon with shorthand terms. The need for efficiency is also why operators must prioritize communication. Important information should be communicated with higher priority. Operators need to know when to speak and when to listen to allow the important information to be transmitted and received efficiently.

B. Communication During Organizational Change

For every process of managing change—which happens often in modern organizations—a communication plan is vitally important. The managers need people to support the change effort, and the people need to understand what changes are coming and what the managers expect the people to do.

The managers of the change effort should develop a strategy for communications, which should include a plan, schedule, messages, and a process for soliciting and receiving feedback.

For audiences at every level of the organization, managers of the change process should develop unique messages relevant to that group. The message must contain information that explains the value of the change, including rationale for why the change is needed. If people understand the value of the change, they won't view the change as merely another "initiative" but, rather, as something needed to help them be successful. The message should be prepared by working with audiences before delivery to determine what the audience members need to hear.

Reception of the message is the objective of the communication. What is heard and subsequently understood is much more important than how it is transmitted. Managers should develop a feedback mechanism for understanding how audiences at every level in the organization are receiving the messages. Based on the data received, the managers may need to revise, refine, redirect, and recommunicate the message. In any event, the managers should repeat the message for those who didn't receive it the first time.

The managers should communicate as one united team in the organization. Any indications of a nonunited change effort will be destructive.

People in the organization need to go through the stages of being informed, understanding, accepting, and, finally, engaging the change. A demonstration of supportive behaviors from managers, as the people progress through the stages, is much more powerful than the words spoken.

Multiple communications media should be used. Multiple informal networks and influence leaders should be used. Managers who are communicating the change need to be inspirational. The best leaders practice techniques of communicating to improve. Some techniques that may help when communicating messages to audiences are:

- Start and finish the message with the important concept you want the audience to receive, in a consistent way.
  - o   The manager should include the high-level concept, which is given to set context.
- Make the core content relevant to the individuals in the audience.
  - o   The manager should tailor concepts for each of the audience groups.
- Include information to help answer the following general questions people have:
  - o   "How can I help? What do you want me to accomplish? Do you have preferences on how I should do it, or do you want me to determine the best methods?"
  - o   "What will happen and when?"
  - o   "What does this mean for me in my job?"
  - o   "What can we expect from you (our managers) to support us?"
- Personalize the content with examples.
  - o   People respond well to references to themselves or their coworkers, especially when the references are positive.
- Keep messages relevant and clear, especially if the concepts are complex.
  - o   The message doesn't need to be simple, but it must be made clear.

Here is an example of a poor high-level message I have heard: "We will achieve sector leadership through strengthening our portfolio and sustainably driving efficiency." The two problems with this statement are:

1. This phrase is not relevant to the front-line personnel and their work in the organization, and
2. Members in the audience will not easily understand the precise meaning of each term and the activities intended to achieve the desired result.

C. The Use of Humor When Communicating

Humor can be helpful but only if used in a positive way. Be uplifting. Motivate the audience or make them feel good or comfortable. Never use humor at the expense of others, even with innocent intentions.

Embrace inclusion. Bring the audience in on the joke; never use insider's humor the audience can't follow or doesn't understand. Do not use humor directed at peers in the management level; the audience will see this as competition, and members of the organization will adopt competitive behaviors across organizational divisions.

Stay humble. Never use humor that appears to boost your own ego.

Create relevance. Every comment, including humor, should be relevant; if it isn't, don't use it. Don't use humor to fill space.

D. The Minefields of Humor

To be effective communicators, managers must understand and avoid the *Minefields of Humor*. Know that most people intuitively sense the underlying truth in humor. Some communicators use humor for the following purposes, which detracts from their message:

- Attempts to mask low confidence.
  - o Humor is often used to mask low confidence or when the speaker is nervous. Unfortunately for the speaker, this humor is counterproductive and highlights the insecurity. Audiences easily identify the different styles of humor used by various speakers with low, balanced, or excessively high confidence.
  - o *Distraction humor* is used when equipment fails during the presentation, such as sarcastically stating, "The projectionist had one job to do . . ."; this is perceived as distasteful to listeners who generally will sympathize with the victim.
- Attempts to inflate own ego at the expense of others.
  - o Again, audiences easily pick up on self-aggrandizing. I have been in rooms, including twice in the White House, when executives and senior managers have engaged in this kind of behavior, which makes the audience uncomfortable.

- Attempts to reduce social risk to the speaker.
    - o A communicator uses this type of humor to say something controversial, often to a listener who is higher in the chain of command. If the statement backfires (is rejected by the listeners), the humor is intended to be a safety net as it implies, "I was just kidding." Nevertheless, the astute listener easily senses the underlying controversy in the statement.

E. The Minefields of Disagreement
Communicators must understand and avoid the *Minefields of Disagreement*. Disagreement expressed by a leader or person in a superior position in the hierarchy creates the risk of shutting down valuable conversation.

- In a group setting, the speaker should avoid saying, "I disagree with you," especially if the speaker is a powerful and influential leader of the members in the audience.
    - o This exchange makes the disagreement feel personal and may have negative emotional or social consequences for the targeted individual and other members of the audience.
    - o This disagreeable statement may inhibit valuable conversation in the future from people who hear this exchange.
    - o This disagreeable statement can alienate the speaker, as listeners in the group may sympathize and side with the individual who is being targeted by the speaker.
- To express disagreement, while promoting a healthy culture of operating excellence, the speaker should:
    - o Attempt to understand another's point of view.
    - o Acknowledge the value of another's point of view.
    - o Use phases such as, "I understand and value your opinion, as viewed from your perspective; I view this from another dimension [explain]; thank you for sharing your perspective; please continue to share in the future."

Communication used to support operations and change management must be effective, which implies accurate. The lexicon must be precise to eliminate ambiguity and uncertainty. If the operations are time-critical, the communication must be efficient.

Information must be conveyed well. But the ultimate goal of communication is the accurate receipt and understanding of the message. The communicators are responsible for verifying the listeners accurately understood the information.

18. Be Prepared Mentally

Dangerous operations can create debilitating psychological pressure in operators. An avalanche of negative thoughts about failure, injury, or death can overrun operators who enter hazardous environments with inappropriate or counterproductive mental attitudes. When an operator succumbs to such thoughts and dwells on the possible negative outcomes, that operator may no longer be able to assess the environment and the situation effectively, make correct decisions, and take appropriate actions. This compounds the danger. The operator enters a death spiral in the mind.

Remember, most of the thirty *Techniques for Operating Excellence* involve each of three characteristics, knowledge, skills, and attitudes. This technique, *Be Prepared Mentally*, is entirely about maintaining the right attitude before and during the operation. The technique is intended to influence the operator to develop and maintain the most appropriate mental attitude in the chosen profession that will enhance the ability to sense conditions, process information, think clearly, assess accurately, make decisions, and operate successfully in any situation with increasing hazards.

The attitude that works best to resist the psychological pressure and avoid cognitive incapacitation can be different for different people in different situations. Understanding which attitude is appropriate requires anticipating the operational situation and relying on previous experience in progressively challenging training or operational situations.

Example: Stage Fright

The example of stage fright, common to many people, may help to illustrate the power of the technique to be prepared mentally, and what can happen if the operator is not prepared sufficiently. As an active astronaut, I never suffered from a lack of confidence when I was on a stage delivering speeches about the space program or talking to audiences, large or small, in elementary schools or the Oval Office. I was comfortable speaking with everyone.

My comfort level changed dramatically one year after I flew my final flight and officially became a "former astronaut" and was speaking about topics other than space flight. During two separate presentations to small audiences in a conference room setting, I experienced a lack of confidence in the material I was presenting, specifically because I was not prepared and didn't maintain a proper mental attitude. Fortunately, the automatic processing in my mind continued to control the formation and delivery of the words I was saying. Unfortunately, the controlled processing in my mind (what I was really thinking about) was suddenly overcome with negative thoughts about what the audience might be thinking of me, the possibility of saying something they disagreed with, and my perceived certainty of failing spectacularly as a public speaker.

These negative thoughts and overpowering emotions lasted only a few minutes, but the psychological damage lasted for a few years. Every time I took the stage to deliver a presentation, my emotionally scarred amygdala responded with a nearly identical flood of chemicals into my brain that caused oppressive negative thoughts about my ability to perform well for the audience.

Audiences almost always gave positive feedback for my presentation performances. Though the audiences never acknowledged they thought my performances suffered, there was a lot of suffering going on between my ears. I knew the problem was in my head. After several years of trial and error, I was able to develop and maintain a particular attitude just prior to and during my presentations that helped me overcome any negative thoughts or emotions that might be trying to surface. I tell myself the audience members are the only people in the room that matter. I am irrelevant. Whether I fail is irrelevant. Helping the audience receive a good experience is the only objective that matters. Five minutes before I pick up the microphone, that becomes my mission—I dedicate myself to giving the audience the good performance they deserve.

Not coincidentally, what works for me on a stage is the same technique that worked for me when I was commanding missions in space. The same technique works for other leaders in hazardous environments. They take themselves out the picture. They don't focus on what may happen to them. Through conscious mental effort and active volitional control, they focus only on the mission, their team, the external environment, and any hazards or obstacles. With their senses tuned outward, they assess quickly and accurately, make the right decisions, and take the appropriate actions. Their brains are unencumbered by any thoughts of self and are operating at peak effectiveness for controlling risk and improving performance.

Before attempting dangerous or critical operations, the best operators psych themselves up. Athletes perform a similar pregame ritual when they visualize their performance to get their minds in an appropriate frame for success.

Some operators in high-stress situations have reported sensations of bradypsychia or tachypsychia, in which the mental processes slow down or speed up and time seems distorted. Being prepared mentally will mitigate the deleterious effects of these psychological conditions.

Find the right attitude that works for you in your operating situations, whether the hazards are large or small. Before entering the hazardous situation, be prepared mentally. Based on previous training, use active volitional control and directed focus of attention to maintain the appropriate mental attitude and allow your brain to perform at peak effectiveness to achieve mission success.

Example: Integrated Training in the Shuttle Mission Simulator
One of the outstanding strengths of the NASA training program for astronauts is the simulator. Because we don't conduct training missions in space, hundreds of hours are spent in high-fidelity simulators on Earth before each mission. Photo 113 shows the crew of STS-86 after one of our final simulations before launch. The simulated environment and the operating scenarios are so realistic the astronauts are able to develop effective skills and appropriate mental attitudes necessary to be successful on launch day, even though some had yet to experience 7.7 million pounds of thrust into their seats.

On my first flight, about twenty seconds after liftoff, a comforting thought flashed through my mind. For an instant, I had a few spare brain cells that suddenly realized, "This is exactly like the simulator." That brief thought gave me vast confidence as I realized I was able to control my mind in exactly the way I had prepared.

Photo 113: The flight crew of STS-86 has completed one of the final simulations at the Johnson Space Center, days before launch (Photo Credit: NASA)

Photo 114: View from pilot's window, STS-113, on outer glideslope for Runway
33, Kennedy Space Center, Florida (Photo Credit: NASA video capture)

Example: In-flight Landing Simulation and Mental Preparation Before Entry Day
Photo 114 shows the view of Runway 33 at the Shuttle Landing Facility from the
right side of the cockpit on STS-113. An interesting mental phenomenon occurs
during reentry into the earth's atmosphere down to the approach and landing.
Early in the entry profile, just after the deorbit burn over Australia, the orbiter is
traveling at 17,500 mph. From an operational perspective, not much is occurring,
so the astronauts' brains can process information relatively slowly. The deeper we fall
into the atmosphere, the slower the orbiter travels, but the operations become more
dynamic. The astronauts' tasks for guidance, navigation, and control are increasing
in number and frequency, so the CPU speed of the brain must increase dramatically.
The slower our vehicle's speed, the faster we need to think. If we delay any single
mental task, the time problem compounds, and we fall further behind in the timeline
of our procedures and mental processes. The only way I found to stay on track
temporally was to anticipate the time problem and force my brain to think faster, as
we descended lower and flew slower.

Before my first landing as a Space Shuttle commander, I heard some astronauts
experienced symptoms similar to bradypsychia on the approach to landing. They
rolled out on final, and the next thing they knew they had landed and time
had passed apparently more quickly than normal. To reduce my likelihood of
experiencing this phenomenon, I prepared myself mentally before reentry. While
on orbit, on the day before landing, I rehearsed the mental processes required
during reentry, approach, and landing, using a laptop simulator. This mental
rehearsal allowed me to accelerate my awareness and cognitive speed consciously.
With this mental preparation before each of my five landings, I experienced no
sense of time dilation or expansion.

19. Be Mindful During Operations

This technique, *Be Mindful During Operations*, is wide-ranging and powerful for operators who are engaged in dangerous operations. Mindfulness is one of the most important skills to learn, develop, and continually practice. Operators use mental discipline and employ *Technical Knowledge, Teamwork, T-0 Vigilance, Cognition,* and *Fields of Vision* to enhance mindfulness. I will explain each of these five aspects and why each is important for enhancing operating excellence, increasing productivity, and preventing accidents.

A. Technical Knowledge

Knowledge of the systems is mandatory to control risk and achieve successful operations. Operators should know everything they can about the systems they operate. With the proper technical knowledge, operators can sense when the performance of a system is beginning to degrade. They can identify failures more quickly and devise corrective actions to rectify the failures more effectively.

With superior technical knowledge, operators can reduce the consequences of errors and prevent accidents. Most importantly, technical knowledge can help operators make the right decisions and take the right actions to improve performance and accomplish missions safely and effectively.

Many operating organizations don't have the capacity to teach their operators all the technical knowledge needed to achieve the best possible results and prevent all accidents. Operators in dangerous businesses have a responsibility to learn everything they can from their organization and then go further in acquiring knowledge. Learning should be a career-long journey.

When I was selected at NASA in 1984, I expected the organization would provide great training. But I also expected this training would not satisfy my desire to learn everything meaningful about the complex operations of the Space Shuttle. I was motivated to study more after I learned, under certain failure scenarios, a pilot's single incorrect action in the cockpit could result in destroying the orbiter.

Early in my naval career, I learned that some of the best sources of technical information were found in the minds of the acknowledged experts. Even better, I found that all the experts I met were willing, and often eager, to share their knowledge. All I had to do was ask nicely.

I began to seek out the best in the Astronaut Office. The Space Shuttle main engines were true wonders of modern technology. They were the most efficient aerospace engines ever made. And they induced fear. Photo 115 is a picture of my wife and me on a launch pad tour looking at the engineering of the nozzle design. Many other complicated systems in the Space Shuttle were just as deadly if operated with insufficient knowledge and skill and an improper attitude.

The following is a short list of a much larger group of astronauts who helped me develop the proper technical knowledge to control the various systems with operating excellence.

Photo 115: Three Space Shuttle main engines, pad tour before STS-86

- John Blaha was the acknowledged expert on the intricate, half-million-pound thrust main engines. He was brilliant, and he shared his knowledge for hours.
- I learned everything I could about the electrical power system from Judy Resnick, who knew more about this system than any other astronaut. She graciously devoted hours to share her knowledge, skill, and attitude.
- Steve Hawley taught me to identify failures without referencing the electronic displays, a useful skill when the displays were not available. His brain seemed to communicate with the flight computers in ones and zeros; some things I couldn't learn from him, but he was always willing to share.
- Hoot Gibson was, far and away, the best stick-and-rudder pilot I have ever met. After a mission, a commander has only one chance to land the one-hundred-ton glider with the lift-to-drag ratio of a brick piano, so I asked him to explain how he did it.
- Mike Smith, before he died with Judy on *Challenger*, taught me how to organize the information I was learning. A few years later, NASA transformed the notebook he taught me to develop into the one-thousand-page *Shuttle Crew Operations Manual*.
- For the mental aspects of making command decisions during spaceflight, the best ever was Dan Brandenstein. Fortunately for me, I had the privilege of flying under his command on my rookie flight, so I tapped his brain for all the knowledge I could acquire during one year in training and two weeks in space.

The crewmembers in the office were not the only ones willing to share information. The entry guidance displays were not intuitive for a pilot, so I found the engineer who designed them. Larry McWhorter was a great source

Photo 116: Video capture of flight-deck crew approaching the deorbit burn, *Endeavour*, STS-113 (Photo Credit: NASA video capture)

of knowledge. Without invitation, I politely knocked on his door, introduced myself, and asked if he could help me understand his displays. He graciously gave me his time and provided invaluable information that helped me on six hypersonic reentries into the earth's atmosphere. (Photo 116 shows the displays in the cockpit.)

The best training is already close to you. Learn from the best. All you have to do is ask. The willingness to share information is a common characteristic of great operators and engineers in dangerous occupations.

B. Teamwork

Teamwork starts with a critical self-assessment by each member of the team. Individuals must know their own strengths and weaknesses before understanding how they can and should contribute to a team's success. In the Astronaut Office, training begins with solo training. Before being assigned to a crew, astronauts must acquire the technical knowledge and skills to operate complex systems. Just as importantly during initial training, evaluators and managers continually assess how well individual astronauts demonstrate the proper mental attitude and interpersonal skills necessary to be contributing members of a team.

Next, prospective crewmembers are assigned to train collectively in nonspecific flight training to learn how to work well on a team. Eventually, if individual astronauts demonstrate the proper teamwork behaviors, they are assigned as members of a crew on a particular mission. Operators become intimately familiar with the strengths and weaknesses of their teammates and know how to contribute to the success of the whole team. The best teams are populated with operators who know how to achieve the best results by developing the most effective methods for working together to maximize strengths and minimize weaknesses. The best operators continue to perform analyses of their personal error tendencies and develop mitigation strategies, which often involve the support of teammates, to minimize the adverse consequences of personal errors.

The elite and highest-performing teams can be distinguished by another characteristic often overlooked by lower-performing teams. The most successful teammates maintain an attitude of *selflessness*. Their behaviors indicate they are more interested in the success of the mission and helping teammates than they are in personal recognition for their own contributions. Experienced observers will notice individual personalities and solo performances begin to disappear in teams with the highest level of teamwork and the greatest achievements.

High-performance teams display another distinguishing characteristic. In conversations or debates before operations, members from high-performance teams listen intently to the opposing viewpoints of teammates. At the appropriate time, the listening crewmembers ask questions for clarification or probe deeper to understand their teammates' opinions or positions more clearly. They are less interested in presenting their own opinions and more interested in understanding why their teammates think the way they do.

Members from low-performance teams with average teammates will be the ones who are reloading as they listen to an opposing point of view, while they wait for their chance to "fire at will" and present their own opinions.

Elite teammates understand how other teammates think, and they make the best decisions based on their collective wisdom. The team's decisions transcend the opinions of any single member and are made better through a deep understanding of all points of view.

C. T–0 Vigilance

T-0 Vigilance is the term I use to describe the hyperattentive, focused, disciplined, mindful, and intensely effective way astronauts sense conditions, process information, and assess situations before and after liftoff. (The countdown clock is anchored at time "T minus Zero seconds," which occurs at the moment of solid rocket booster ignition.) During critical phases of operations, elite operators in many industries— including military, maritime, aviation, space, energy, athletics, and medicine—are *present*; they stay in the moment and focus on everything necessary to be successful. *Be here, now.*

Through intensive training, vast experience, and mentoring, these operators have tuned their senses, and they know exactly what to look for when assessing a situation. With the skill to filter out all extraneous information and distractions, their perceptions are heightened, and they sense every relevant aspect of the operation. They miss nothing. They see minor fluctuations in small instruments or gauges in their peripheral vision. They feel slight differences in thrust vectoring of the engines or minor increases in vibration of equipment. They hear the voices of crewmembers and the ground team and are sensitive to changes in thinking or emotional states. They sense subtle changes in the environment, the cockpit, the operating theater, the forward operating base, the playing field, and the battlefield.

Fear is a useful emotion that can motivate an operator to study, practice, and prepare for high-risk operations. Worry is unproductive fear and becomes a consuming emotion when nothing is done to eliminate it. Long before the pressure of danger builds, worry must be eradicated and fear must be channeled toward preparation. During high-risk operations, fear must be managed and repressed through active volition and cognitive discipline.

During the dangerous and dynamic phases of high-risk launch operations when events happen quickly and minor failures can cascade to catastrophe, the best astronauts do not worry about anything in the future more than ten seconds from the present. Nothing else matters past this temporal horizon. Because they have greater mental clarity in the present, good crewmembers have the ability to sense everything, assess all possible situations, make effective decisions, and respond correctly to any emergency. If a crewmember can develop greater mental clarity under seven and a half million pounds of thrust, that mindful crewmember will be a successful astronaut.

The T-0 Vigilance aspect of this technique, Be Mindful During Operations, is mastered only through intensive and directed training, dedicated practice in progressively more demanding situations, and disciplined mental preparation before the operation. These investments in training and preparation will be valuable. With heightened vigilance in the present, mindful operators are able to prevent accidents and improve performance in any dangerous operation. Mastery of T-0 Vigilance is enabled through purposeful use of the next aspect, *Cognition (Controlled and Automatic)*.

D. Cognition (Controlled and Automatic)

By understanding and using two different mental processes, we can operate more effectively in dangerous environments. The brain can be trained to use the high-level cognitive process of concentrating on a single challenging task, such as preventing accidents while driving in hazardous conditions. And, we can use a second, less focused, mental process to perform several tasks simultaneously.

These two kinds of cognitive processes are called *Controlled* and *Automatic*. We are mostly aware of our brain's Controlled, or upper-level, executive process. The contents in that process are what we are concentrating on or thinking about at any given time. Most of the brain's processing is accomplished in a different mode, the lower-level Automatic mode. I rarely think about the mechanics of playing my drums on stage (Automatic processing), but I do think about where my bandleader is trying to take the audience in our music (Controlled processing).

As the name implies, we can't control the Automatic processing in our brains. But, having Automatic be automatic is a good feature. We want this mode to operate on its own so it will move us quickly out of danger to save our lives. Automatic also allows us to breathe, sweat, walk, and chew gum, all at the same time. Though we can't control Automatic, we can train it to perform better through practice. That is the primary purpose of training and the value of becoming skilled.

Sometimes, we feel like we don't have much control over Controlled. Thoughts spontaneously appear. Others disappear, never to be heard from again. But, we have more control over Controlled than we might think. The most obvious type of control is choosing what to put into Controlled. We can decide to think about a red ball, if we desire. We can decide to concentrate on our driving tasks while driving, if we want to stay alive.

With training, we can master two other types of control over Controlled, which are particularly useful for operators in dangerous environments.

First, we can create a rule-based or goal-activated mental subroutine, which is preset to select the thoughts that go into the Controlled process (sort of automatically, after the subroutine is set). For example, without knowing ahead of time which items to think about specifically, I can decide to notice and think about any item in this room that is colored red. This is useful in a refinery's control room where the displays will show a red number if an instrument senses a violation of a redline limit.

Or, we can decide to notice and think about any hazard that can kill us on a factory floor or the flight deck of an aircraft carrier. Again, this is useful if I'm an operator controlling complicated machinery or flying a high-performance jet and trying to predict and prevent potential accidents when the exact nature of the potential accident is unknown. To take advantage of this mental ability was exactly the reason I created each of these *Techniques for Operating Excellence*. These techniques are mental subroutines operators can insert into the Controlled processing in their minds to achieve operating excellence, improve performance, and be able to identify and prevent potential accidents coming from the unknown.

Second, we can do the opposite. We can set up a preconditioning filter that keeps unwanted ideas out of our mind's Controlled processing. I don't agree with the common notion that I will inevitably think about a pink elephant if someone tells me not to think of a pink elephant. Through active volitional effort and directed focus of attention, I can think about something else and prevent the unproductive thought from intruding.

When I'm on the launch pad, I can choose not to think about the prospect of dying twenty minutes later, when the rocket ignites. By concentrating on the present and conducting my prelaunch operations professionally with operating excellence, I can repress the fear about the future.

When driving home after a bad day at the office, I can choose not to think about my arrogant boss and, instead, focus on my driving tasks so I can stay alive (to face him or her again another day). Operators need the ability to compartmentalize by using disciplined concentration and focus of attention through active volitional control over their mental processes.

Choosing what goes into the Controlled mode in specific situations can save lives. Dangerously for you, many drivers will be operating their cars around you by using only their lower-level, Automatic mental processing for their driving tasks, which has much less ability to sense the weak signals of an impending accident. But you can choose to put your driving tasks into your Controlled mode and keep the other drivers from killing you.

The Controlled mode has limitations. First, it's slow, compared with Automatic. Second, although the Controlled mode is highly capable and effective, it can handle only one task at a time. To conduct multiple tasks simultaneously (to multitask) we must use our lower-level, Automatic processing. That's why pilots conduct extensive training to become skilled in operational tasks. Once skilled, the tasks are relegated by the brain down to the Automatic mode. Training also helps operators learn to switch rapidly in their Controlled mode between separate tasks under sequential conscious control.

The result of being able to handle multiple tasks in Automatic, and switch quickly between serial tasks in Controlled, gives us the illusion of multitasking. Though we are *conducting* multiple tasks simultaneously, we're really consciously *controlling* only one task at a time. This is an effective way to perform in high-tempo operations, and it is why pilots, athletes, surgeons, SWAT team members, firefighters, musicians, and many other operators train so much. Intensive training, in progressively more demanding situations, gives humans the ability to perform more tasks simultaneously in the Automatic mode, while switching attention rapidly and effectively in the Controlled mode as the situation demands. For a wonderful treatise on multitasking, see *The Multitasking Myth* (Ashgate, 2009).

The Automatic processing mode has limitations, too. As we conduct tasks in Automatic, without thinking, we have little ability to respond to unforeseen or untrained situations and modify our performance of those tasks. Also, although our brain's response time (our reaction time) is quick in the Automatic mode, the response is only triggered if an external stimulus grows large enough to penetrate our senses.

If a motor vehicle driver is operating in the Automatic mode, that is, *driving on autopilot*, the initially weak signals of an impending car accident may not be sensed until it is too late to prevent the accident. This is why we should not drive while using a cell phone. The brain is using its adaptable and highly sensing Controlled mode exclusively for the conversation on the phone. The driving tasks remain in the lower-level, Automatic mode of processing, which only senses signals that are strong enough to break through. Total response time to prevent an impending accident will be reduced because the Automatic mode has limited ability to sense weak signals. That precious second of lost reaction time while talking on a cell phone may be the last second the brain ever experiences.

When we drive a car, we should keep our driving tasks in the Controlled mode of mental processing. We should concentrate on the dynamically changing situation and use our heightened ability to sense weak signals to prevent accidents. In the oil-and-gas industry, when board operators are issuing commands from a control center to lift gas in a well, the board operators should concentrate on the lifting process. When field operators are in a high-hazard environment, they should concentrate on searching for vulnerabilities and identifying hazards.

Example: Using a T-38 for Spaceflight Readiness Training
Astronauts use Spaceflight Readiness Training flights in airplanes to improve their cognitive skills in both the Controlled and Automatic mental processing modes. Photo 117 shows my crewmate, Vladimir Titov, on a training flight in the backseat of our T-38, with another airplane on our wing.

Photo 117: Russian Cosmonaut Vladimir Titov in the T-38
for Spaceflight Readiness Training (Photo Credit: V. Titov)

When pilots first learn to fly, they accomplish single tasks with directed attention and conscious thought in the Controlled mode of mental processing. Sequentially, they move on to conduct another task, but, again, they must consciously think about performing that next task.

The goal of flight training is to practice each task until it becomes a skill, which means the task can be accomplished automatically without thinking. The brain unconsciously conducts the tasks in the Automatic mode of mental processing. Accomplishing tasks automatically is what athletes mean when they say they try to not let their brain get in the way; they perform better when they react automatically. The more we practice in airplanes, the more we can relegate individual flying tasks from the Controlled mode down into the Automatic mode.

We also have to know when to bring certain tasks back up into the Controlled mode for conscious processing. For example, when we are flying formation on the leader's wing at night in the clouds, we usually concentrate on maintaining precise position relative to the leader's airplane (Controlled). Periodically, we consciously divert our attention to navigation, our engine instruments, and our fuel state to verify our safety (also Controlled). During those periodic times, the formation flying, even though accomplished through manual control inputs, is performed without thinking (Automatic).

So, there are two purposes for spaceflight training. First, in basic training, we want to become highly skilled so we can relegate as many tasks as possible into Automatic processing, which allows us to perform many tasks simultaneously without conscious control. Second, in advanced training, we develop the valuable mental ability to bring selected items from Automatic back up to the Controlled mode to create periodic focused attention on critical aspects of the operation while anticipating changes.

This second purpose, often overlooked in other organizational training programs, is critical before a spaceflight. We must learn to bring items back up from Automatic to Controlled for devoted cognitive attention. We practice rapidly switching between one important task and the next, using the Controlled mode—and we learn what tasks to switch to and when to switch. If we have an emergency, devoting attention to the emergency and making decisions are simple tasks. But what if the engine's performance is starting to degrade minutes before complete failure? How will we notice slightly degraded performance if we aren't looking for it, while functioning on our brain's autopilot?

We have to know what to look for and when to look. If we can't perform this second skill of bringing selected items back to Controlled, we will never see the next accident coming and we won't be able to prevent disaster. I knew my training was effective when I began to develop the ability to identify unknown failures in the Space Shuttle simulator before our on-board computers annunciated the failures. To succeed in the simulator and survive in real-world dangerous environments, we had to know what to look for—and when, where, and how to look—while the potential accident was not yet demanding our mental attention.

That's why Titov and I practiced so much before spaceflight. We needed the ability to switch tasks back and forth between Controlled and Automatic, to sense and process information, make decisions, take actions in dynamic environments and hazardous situations, and predict and prevent impending accidents before they killed us.

Example: Cars on the Wrong Side of a Railroad Barrier
The city of Houston has a light metro train system. For more than a decade, multiple accidents have occurred every year between automobiles and the metro train that runs on tracks through the city. Look at Photo 118. Do you think the drivers, who are stopped on the wrong side of the railroad barrier in the picture, were using their Controlled mode or Automatic mode for their driving task? Should they have prevented this situation? How could they have prevented this situation?

Photo 118: Were these drivers operating under Controlled or Automatic processing?

E. Fields of Vision
After the equipment and systems have been designed, the training has been completed, and the operation commences, an operator's mental discipline becomes the most powerful instrument for reducing errors, avoiding hazards, and improving performance. The *Fields of Vision* through which an operator perceives can be expanded or contracted through active volition to suit the particular conditions of the dangerous operation. During critical or complicated tasks, focused attention may be warranted. In dynamically changing or especially hazardous situations, having a

wider perspective to increase sensory inputs may be better. The ability to concentrate is valuable in both cases. Training can help the operators know when to focus on the complex tasks at hand and when to open mental awareness and maintain a wider field of attention to prevent accidents.

I began to appreciate the power of Fields of Vision and teamwork on my first flight in space with Dan Brandenstein. If he went head-down to focus on the details of a critical flying task during proximity operations near a massive satellite, I had to be head-up maintaining the bigger picture, scanning the operational horizon for impending critical events and any emergency that might be coming at us sideways. Conversely, I was allowed to focus my attention on a detailed procedure only when he was maintaining the wider perspective.

If T-0 Vigilance, or intense mindfulness in the present, is described as, *be here, now*, then Fields of Vision is, *be where I'm not*. Always maintain the perspective complementary to that of your teammates, wide or focused.

Example: A-7 Emergency Shutdown on a Carrier's Flight Deck

Photo 119 shows the congested area on the flight deck of an aircraft carrier. You can see from the picture the flight deck personnel sometimes work near the engine intake, under the nose of the A-7. Early in my career, I surmised that someday, someone would wander unintentionally too close and be sucked into my intake. I could see this as inevitable from assessing the behaviors of the personnel.

During the engine start and prelaunch operational checks, my attention was necessarily focused inside the cockpit on the displays and switches, anticipating aircraft failures requiring immediate corrective actions. Yet, I knew the day or night was coming when I would be sickened by the sound of my compressor blades chewing up an inattentive deckhand who got ingested.

Photo 119: A-7 Corsair aircraft on the flight deck of the USS *John F Kennedy*

So, I trained myself to devote simultaneous attention outside my cockpit to every person who came near my airplane after I started my engine. I also knew, from observing the hectic activities of the personnel, I would have no time to assess the situation and decide to shut down. In my mind, I visualized a minimum distance "shutdown zone" around my intake and modified that distance according to the speed and direction of anyone who might potentially violate my limits. This is the point of the technique. As an operator in a dangerous business, I decided to train my brain to use different simultaneous fields of vision for attention. With this cognitive skill of being more mindful of the situation around me with people near my hazardous engine, I could apply a predetermined decision algorithm for automatic shutdown.

In nine months of operations on aircraft carrier flight decks, I shut down my engine on six different occasions when sailors got too close. As I had anticipated, I had no time to think. I was glad I created the mental algorithm so my shutdown decision was automatic. The sailors were also glad. Three of the six took the time to find me three hours later, after I returned and landed on the ship, to thank me for shutting down and saving their lives. None of them were aware that they were too close until they heard my engine spooling down right next to their helmet. On all six occasions the air boss in the tower became irate and chastised me over the radio for screwing up his launch sequence, since I needed extra time to re-start my engine. None of that mattered to me. I only cared that six sailors didn't die.

The best operators employ technical knowledge, teamwork, T-0 vigilance, cognition, and fields of vision to enhance mindfulness during operations. They know when it is time to concentrate. They know *how* to stay in the moment, and how to selectively use both the Controlled and Automatic modes of cognition and the appropriate fields of visions to predict and prevent accidents and improve performance.

20. Think Fast and Act Deliberately

Through personal experience and learning from the performance and mistakes of others, the best operators learn to *Think Fast and Act Deliberately*. In dangerous operations, where split-second decisions can save a live or end a life, the best operators must think fast, yet not act too fast. Operators must assess the situation and process information quickly but must not rush and make mistakes. It takes a certain kind of person with specific mental skills to be successful in fast-paced operations involving dynamic, constantly changing situations that can degrade to catastrophe through inappropriate action or passive inaction.

Good operators sense conditions, analyze options, and develop solutions quickly, but implement corrective actions deliberately. Speed is desired, but precision is required. It's good to be fast, yet operators can't afford to be wrong. Compared with machines, this is where people excel. The human mind surpasses the capabilities of computers in judgment, intuition, and decision-making when given incomplete information. People with experience can determine solutions quickly, but they have the self-awareness to know they can make mistakes. This is why we will always need operators at the controls. The will to live drives an elite operator to be accurate in the face of danger. With years of experience and mentoring, the best learn to think fast yet act deliberately and purposefully.

Example: Bill Shepherd and Sergei Krikalev, Veteran Space Travellers

The quickest way to get in trouble in space is to act quickly. It's too easy to make mistakes, and some mistakes are irreversible. When astronauts are piloting a rocket, there is a tendency to believe we must act quickly because we are moving so fast. The consequences of emergency situations are much greater in space, so we have a feeling we must act quickly to prevent escalation of the danger. Actually, the opposite is true. Because we are moving so fast and because the consequences of emergencies are greater, we must be more accurate.

Operating excellence comes from being deliberate, not being fast. In space, distinguishing the rookies from the veterans is easy. The first-time flyers usually moved much more quickly through the cabin. Often, they left a trailing wake of

Photo 120: Russian Cosmonaut Sergei Krikalev, ISS-01, with STS-102 arriving (Photo Credit: NASA)

debris, as they kicked books, equipment, and experiments on their way to perform a task in a hurry. Their inefficiency usually put them further behind in the timeline, which caused them to rush even more. The experienced flyers looked as if they were moving in slow motion. They were extremely efficient and effective. There was no wasted motion. Go slow to go fast.

The best operators I ever saw in space were Captain Bill Shepherd, US Navy (Ret.), and Russian Cosmonaut Sergei Konstantinovich Krikalev (photo 120). A former Navy SEAL, Shep commissioned the International Space Station as Commander, Expedition-One. Sergei was his pilot on Expedition-One, and later was Commander, Expedition-

Eleven. Shep and Sergei were a joy to watch in space. The way they moved seemed magical. The way thought and operated was inspirational. Shep and Sergei had a cool steadiness in their thought processes and decision making that enabled operating excellence.

In a dangerous business, underwater, on the ground, in the air, or in space, elite operators perform calmly—think fast and act deliberately.

21. Recognize Divergence

The best operators are highly skilled in identifying the early signals of an impending accident and taking corrective actions to prevent the accident. But, what are the signals, and how can operators learn to identify them? In some organizations, personnel have popularized the phrase "identify weak signals" yet offer little explanation of the meaning to help operators understand what to look for.

In my experience, accidents don't occur after an expected sequence of normal events. Generally, accidents happen only after the sequence of events has been disrupted or has begun to deviate and diverge from the expected pattern. One or two slightly abnormal events can initiate and perpetuate a deviation from the expected pattern. Often, this slight divergence from normality does not cascade to a greater divergence, resulting in an accident. But sometimes it does. Operators shouldn't rely on historical statistics to save them.

Elite operators *Recognize Divergence*. The best have learned to recognize the earliest and slightest divergence from expected or normal operations. This is what the safety personnel mean by "weak signals." If the smallest divergence from normal operations is recognized as soon as it occurs, the necessary corrective actions to prevent cascading failures will be smaller, easier, and often less costly, to execute.

Accidents are usually preceded by divergent situations. Operators who recognize divergent situations can respond with appropriate corrective actions to prevent impending accidents and improve operating performance.

The following is a list of some indications—and possible causes—of a diverging situation:

- Change in scope of work
- Unexpected event or situation
- Abnormal event or situation
- Minor failure
- Hurrying
- Distraction
- Fatigue
- Psychological or physical stress

In an evil twist, one of the more common indications for a diverging situation is a distraction, which is inherently difficult to recognize. A distraction is a new event that automatically draws the attention of the operator away from whatever was previously in the Controlled mental processing. Distracted operators don't know they are distracted, and they fail to see their situation is beginning to diverge, precisely because they have become distracted.

In a dangerous business, the front-line supervisor should provide training and coaching to help the operators learn the technique and master the skills to Recognize Divergence.

Some operators can develop the ability to think at a metacognitive level, to think about how they are thinking. With practice, an operator can cognitively jump to a higher state of mind, or level of abstraction, and view the situation from the perspective of an imaginary higher-level dispassionate observer. The operator can assess operations and make self-observations about the situation and personal performance, such as, "Interesting; I see the situation is beginning to diverge from the plan and I'm becoming distracted, which may cause further divergence."

During operations, routinely search for indications of diverging situations similar to those listed above. When you recognize one of the indications, compare your current performance relative to your predicted operational plan. If your current performance is beginning to degrade, you have successfully Recognized Divergence.

Operators who have developed an ability to recognize a divergent or deteriorating situation have the greatest chance of averting disaster. Using the technique Recognize Divergence enables operators to be aware of situations that lead to error-prone behavior, such as rushing during a procedure, minor failures in a system, distractions, fatigue, and stress in themselves or other crewmembers. Cognitive lock, or tunnel vision, is a debilitating reduction in the normal mental processes of pilots in flight operations under stress or confusion. With experience and proper training to recognize diverging or distracting situations, cognitive lock is much less likely to occur.

Distractions were the greatest causes of errors during T-38 aircraft operations at NASA. The key to reducing the negative effects of distractions is to recognize the diverging situation early. Understanding what can cause distractions supports anticipating and recognizing diverging situations:

- Distractions can be caused by *normal conversation*;
- When using a checklist a checklist or procedure, any *interruption* experienced, which demands new thoughts or actions, is a potential source of distraction;

- A pilot can experience *local preoccupation* with a complex procedure, a single confusing or critical step in a checklist, a salient event, or a sudden malfunction in a system;
- The operator may experience *global preoccupation* with a personal situation, or a family or external issue.

During every critical phase of flight on all six of my space missions, we experienced distractions. Successfully recognizing these distractions was the first step in preventing degradation in our performance.

For further reading on the concepts of diverging situations and distractions, see *The Multitasking Myth* by Loukopoulos, Dismukes, and Barshi. They present insightful analyses and writing on these challenging operational hazards.

Example: Diverging Operations in Airshows
Aviation has a long, tragic history of accidents occurring during airshows after pilots failed to recognize situations diverging from safe parameters. A common reason for their failure was being distracted by the presence of the crowd. Mishap pilots have performed their demonstration maneuvers perfectly during practice on Thursday and Friday with no crowd, but then have crashed on Saturday or Sunday while they were aware of the crowd. The mishap pilots likely didn't realize how much their situation was diverging from safe parameters. Some of the airshow pilots were reluctant to abandon their maneuver until it was too late to save the plane because they didn't want to look bad in front of their fans.

In 1978, I had an opportunity to participate in an ordnance demonstration airshow that was beginning to diverge from the plan. We were on the shakedown cruise on the USS *Eisenhower*, and President Jimmy Carter and the First Lady were aboard to observe the demonstration. The good news is the two divergent operations only resulted in humorous stories.

First, a flight of four A-7 Corsairs from our sister squadron rolled in and dropped training bombs on a simulated target, indicated by smokes in the water, just off the ship's port side.

Next, a flight of A-6 Intruders from the medium attack squadron rolled in to deliver their ordnance. Unfortunately, they bombed the wrong side of the ship. Fortunately, embarrassment was the only adverse consequence. After we landed, we were told the president was unfazed. The story was told that he stood up slightly, rotated his chair 180°, and sat back down to continue watching the wrong-way airshow.

After the backward bombing, it was our turn. I was flying the number-three position in a division of four A-7s. Our job was to roll in, line up on the smokes, and strafe the sea on the port side of the ship with our 20 mm M61A1 cannons. Here is what happened in front of the spectators as we fired our guns.

The number-two aircraft was a little bit late in pulling off from his run. This meant I had to diverge slightly from the plan and delay firing my gun until he was clear. By then, I was at a lower altitude than planned. This cascaded to a double delay for number four behind me. He saw all this occurring, and he chose to roll in wide so he could commence his firing run at the proper altitude but offset laterally from my firing line.

The result was he began to fire his gun while I was still in my run. To the observers on the deck of the carrier, who couldn't see the lateral offset, number four appeared to be trying to shoot me down. Our divergent operation was conducted safely, but it may not have appeared safe to the president of the United States. In hindsight, after recognizing the plan was beginning to diverge, I should have pulled off at the planned altitude without firing my gun.

Example: Gas Nozzle

Recognizing divergence is incredibly challenging. When you are late and getting further behind, it's hard to recognize you are beginning to rush because your brain is occupied with trying to catch up. You are more error prone. One mistake drags you further behind, so you rush even more. Now, of course, you are even more error prone. You can only break the cycle if you can mentally rise above the situation and recognize you are hurrying.

Notice photo 121. How many times do you fill your gas tank using your mental autopilot while you're thinking of something else? Even if you are conscientious about safety and operating excellence, and ninety-nine times out of one hundred you stay

Photo 121: Distraction can quickly and easily cause a driver to forget to disconnect the gas hose.

mindfully in the moment and you remember to remove the nozzle from your gas tank, one time out of one hundred you will be distracted. When you become distracted, the only thing that will save you from driving away with the nozzle still attached is your Automatic mental process, which you can only hope has automatically removed the nozzle. If the distraction was being late and rushing to catch up, your Automatic processing can be tricked easily into bypassing that step.

The equivalent consequence to a pilot after being caught in the trap of distraction is forgetting to lower the landing gear before landing. The oil-and-gas driller hits the blind shear ram by mistake and drops the pipe in the hole. The nurse administers the wrong dosage of medicine. The mountain climber forgets to secure the harness. The driver runs the red light.

If you think distractions won't cause you to have an accident, someday they will.

Example: Abseilers Recognize Divergence

I once met a supervisor on a team of abseilers who was the best I've ever seen in his ability to Recognize Divergence in his operators. In the oil-and-gas industry, the abseilers are the employees who work at height, hanging from lines, and perform maintenance on the structure of the platforms at sea, high above the waterline. This is one of the more hazardous jobs in a hazardous industry. Photo 122 shows an abseiler in his element, doing what he does best.

As I spoke to the supervisor, I began to understand and appreciate the psychological and emotional readiness his operators needed to stay alive when hanging hundreds of feet above the water. If one of his workers failed to use his harness correctly, that worker may have plummeted to his death. This supervisor felt his responsibility was

Photo 122: Abseilers: the best supervisors Recognize Divergence.

to observe, assess, and evaluate the psychological and emotional readiness of his crewmembers as they came to work each morning. Because he cared so much for each person on his team, he took his responsibility seriously. He met with each worker when they arrived. He observed each of them as they began their workday. If any of his workers showed any signs of distraction, he asked them not to climb that day but, rather, to support the team with other tasks or administrative duties.

Using his leadership skills, this supervisor helped his workers believe that being asked not to climb was not punitive, and it certainly was not a sign of weakness in the worker. Anyone can have a bad day. That's a day you don't climb.

22. Share and Challenge Mental Models

Techniques 22, 23, and 24 are similar. In each, the goal is to communicate effectively with others to improve decision making in dangerous operations that are volatile, uncertain, complex, and ambiguous.

Managers and operators, who are involved in the operations, receive information through their senses during the course of the work. Individually and collectively, they begin to form an understanding of the state of operations in their world, based on sensory information and assumptions informed by experiences. In their minds, they form pictures or models that represent their understanding of the state of operations.

No person can sense everything about the operation at all times. Sensory information is interpreted differently, and people have different experiences that fuel assumptions. No mental model will be completely accurate in representing the actual state of operations.

The goal of this technique is use communication to develop a mental model of the state of operations that is accurate enough to enable managers and operators to control the operations successfully. Each of us should improve the accuracy of our individually held mental model by sharing our understanding with others and integrating feedback. Collectively, through sharing our models and challenging others, the group can improve the accuracy of its mental model of the state of operations.

When flying, the best pilots continually challenge their current mental model of the existing situation. They cross-check their instruments and verify their navigation. These pilots are always thinking about what they might be missing or misinterpreting. They are attentive to the slightest indication that their mental model of the situation might be incorrect. Pilots who routinely challenge their mental model will not be blithely confident they have sufficient altitude at night or in the clouds when in reality they are dangerously close to death by mountaintop.

Example: Sharing Mental Models with the Fight Controllers in the Mission Control Center

Sharing mental models with others was the most important and effective technique we used to eliminate nearly all errors during spaceflight on STS-113. Before beginning every procedure, we reported to the Mission Control team the title of the procedure and the page number of the Flight Data File document (checklist) we were using. We also made a quick radio transmission stating our intentions before every change we made to the autopilot for flight control or maneuver execution. We told the ground team what we were thinking and what actions we would take before every operations sequence transition, water dump initiation, procedural block in the rendezvous plan, and step in the robotic arm operations—really, any action that could have had negative consequences in the outcome of the mission. This enabled the Mission Control teams to be more effective in assisting us during operations, and they helped us eliminate nearly all mistakes in flight.

Photo 123 shows Paul Lockhart and me on the flight deck of *Endeavour* during the postinsertion timeline. We had just arrived on orbit and were still wearing our pressure suits. You can see I was reaching for the microphone to communicate with our ground support team of flight controllers in the Mission Control Center in Houston to share information.

Photo 123: Postinsertion, sharing mental models with MCC, STS-113 (Photo Credit: NASA)

STS-113 was a highly successful crew transfer and assembly mission to the International Space Station. A significant reason for that success was our use of this technique, *Share and Challenge Mental Models*. Before launch, we decided this was the technique that would take us to the highest level of performance. We launched with a crew of seven, yet we expanded our effective crew size to more than one hundred in the way we coordinated our operations with the teams in Mission Control.

At every step in our two-week mission, via radio communications, we kept the controllers involved by sharing our mental model of the state of our operations. Through this process, the flight controllers knew exactly how we were thinking and what we were doing before we thought or did anything. They became an integral and valuable part of our crew. Mostly they confirmed our constantly updating mental model. Occasionally, they challenged our assessment and helped us improve our operations with suggestions or corrections to our plans and actions.

Example: How Mission Control Works with Mental Models
Here's a quick explanation of the inner workings of the Mission Control Center and how the sharing of mental models works. I will use the legendary example of Neil Armstrong's final approach to the first lunar landing on *Apollo 11*. During descent in the lunar module, *Eagle*, the crew received a program alarm, designated 1201, from the onboard flight computer, which was beginning to experience problems.

Before flight, the crew had neither the time to learn nor the capacity to understand thousands of different alarms possible with *Eagle's* computer. But minutes before touchdown on the lunar surface, they needed to understand immediately if the computer was healthy or not and whether to continue the approach or abort the landing.

So, Neil and his copilot, Buzz Aldrin, quickly transmitted the alarm designation to Mission Control in Houston by saying simply, "1201." Gene Kranz was the flight director and responsible for the mission. Gene knew instantly the crew was asking, essentially, "Our computer has a 1201 alarm; are we go to continue or should we abort the approach?"

Gene didn't know the answer, so he quickly asked a member of his team who was responsible for the computer, Guidance Officer Steve Bales. Steve immediately told Gene the crew was go to continue. Clearance to land was quickly transmitted to Neil across 240,000 miles, but that took two seconds at the speed of light. Of course, that was soon enough, and Neil successfully landed on July 20, 1969.

History shows Steve was recognized as the gutsy flight controller who quickly made the right call to enable the first lunar landing. That is appropriate. As the guidance officer in the main control room, Steve was responsible for the call, and he deserves the credit.

But, the history books don't often show who really had the answer to the critical question Neil asked. For that name, we have to continue the story. It turns out Steve didn't know the answer, either. He deferred to his computer specialist in a back room, a brilliant young man named Jack Garman. Before flight, Jack wrote down, by hand, every single alarm that the computer could possibly annunciate. As soon as Jack saw the 1201 program alarm on the downlink, he knew immediately that the computer was experiencing executive overload but that it would not deter the landing.

That's how sharing and challenging mental models works in the spaceflight environment. Neil's mental model had a gap. He asked Gene. Gene asked Steve. Steve didn't know, but he didn't have to ask because Jack had immediately shared his accurate mental model with Steve. Back up the chain it went, all the way to Neil at the moon.

No single person ever has the complete and accurate mental model. Each of us has gaps or inaccuracies. When we share and challenge, the gaps in our mental models are filled in, and the errors are eliminated.

Example: Construction of Mental Models at Three Mile Island
Photo 124 shows the stacks at the Three Mile Island nuclear power plant. On March 28, 1979, the plant experienced a cooling malfunction that caused the core to melt. The reason this accident did not result in total catastrophe was a single supervisor challenged the erroneous mental model the control team held.

Photo 124: Three Mile Island Power Plant

For a little more than two hours, a series of cascading malfunctions caused an increase in heat and pressure in a cooling system, followed by loss of cooling water, which resulted in the nuclear core overheating. As the control team interpreted the indications, assessed the situation in the reactor, and made corrective actions to contain the emergency, they developed an understanding of the configuration of the valves, lines, and tanks in the reactor.

As the emergency progressed, the control team knew a pilot-operated relief valve had opened automatically to relieve increasing pressure in the cooling system. After pressure was relieved, the team commanded this valve to close, and they verified from instrument indications the valve was closed. This action should have restored cooling to the core. The team was convinced the valve had closed.

Their interpretation of this configuration became their mental model of how the reactor was operating. As they understood their model, the valves were now in the proper position, and the core should begin to cool down. But, it wasn't.

That's the problem with mental models. Once our brains develop a model of the world, changing the model becomes exceedingly difficult. Even when our model is inaccurate, we are convinced it is accurate because we watched the model form. That's why we sometimes need other people to challenge our model.

An off-shift supervisor named Brian Mehler entered the control room after the team became confused and was struggling to make sense of the emergency. He had not participated in forming the erroneous mental model. So for him, to challenge the assumptions the control team made was much easier. Additionally, Mehler wasn't on duty yet, so he wasn't responsible for taking any actions. Without the additional psychological stress, Mehler was able to think calmly and postulate which parts of the team's model might be erroneous—the core temperature really was increasing.

So, Mehler built his own model, in reverse. He knew how the reactor must have been configured if the temperature was still increasing. When he compared his mental model to the team's mental model, only two differences existed between their assumed configurations of the reactor. To experience increasing core temperature, at least one of those differences must be incorrect in the team's model and correct in Mehler's model.

His second choice was right. On his suggestion, when the team closed a block valve upstream from the pilot-operated relief valve, which the team believed was already closed, cooling was restored and the major disaster was averted. It turned out the pilot-operated relief valve was stuck partially open because of a mechanical failure. An independent failure in the instrumentation system gave the team an erroneous indication that the valve was closed, misleading them to construct a powerfully convincing, yet incorrect, mental model, which was only broken by the challenge from the off-duty supervisor.

In hazardous operations, even if we are convinced we are right, we should share what we think and ask others to challenge our thinking. Either way, we win. If our thinking was correct, the challenge gives us confidence. If our thinking was incorrect, the challenge may save our lives.

Example: Encourage Questioning and Diversity of Thought

As a leader, when I was faced with an important decision, I was concerned when everyone on the team developed the same opinion during the deliberations. Decisions are always better when one or more team members challenge the conventional thinking of the group.

In a dangerous business, the leader should encourage questioning and diversity of thought. The problem, though, is humans are social. Being part of the group, thinking like the group, and agreeing with the leader feel good. To overcome this tendency for the team members to conform, the best leaders will encourage diverse thinking, show they value diverse thinking, and demonstrate appreciation when they receive inputs from diverse thinking.

The best leaders appreciate having at least one "black sheep in the crowd" (photo 125), who is willing to deviate from conventional thinking and challenge the group. The decisions are always better.

Photo 125: Don't be afraid to be the "black sheep in the crowd" with a different opinion.

23. Challenge "Go" Deliberations

At certain milestones during hazardous operations, managers make important decisions. Usually, a single manager is responsible for making the decision that either takes the operation to the next level or suspends the operation. For shorthand, I will use the aerospace terms "Go" and "No-go" for these two possible outcomes of the deliberation.

Decisions made during hazardous operations in service of important missions can be complex and challenging. Depending upon the importance of the milestone in the operations, often the manager makes the decision after deliberations with a group of people who provide inputs and analyze information. All humans have some degree of bias, based on previous experiences, when they make decisions. A conservative bias toward safety can be advantageous in a dangerous business when astute managers attempt to use good judgment based on valuable mentoring and relevant operational experiences.

But some unconscious biases are counterproductive. Cognitive biases, such as group-think, can adversely affect the outcome of a decision. Individual decision-makers have differing degrees of risk perception and risk propensity, or tolerance to accept risk. Unrecognized biases toward risk acceptance in an already high-hazard operation can destroy programs and kill people.

This technique, *Challenge "Go" Deliberations*, is specifically intended to enhance the quality of decisions associated with risk in hazardous operations. Since Go decisions involve taking the operation to a new level, a Go decision inherently creates more risk than a No-go decision, which holds or suspends the hazardous operation.

Over thirty-five years in operations, I have seen managers passionately debate and challenge No-go inputs much more often than the Go inputs. This emphasis on arguing against slowing down may be caused by the inherent boldness in people involved in space exploration who are motivated to take actions, move forward, and accomplish missions. But these No-go recommendations rarely result in accidents.

I saw much less debating and challenging when a manager or engineer said we can go, we can be more aggressive, and we won't have an accident. Not coincidently, these Go recommendations are the ones that often increase risk and can lead to accidents. Good leaders flip their emphasis on debating and challenging. Wise managers encourage participants in the decision-making process to Challenge "Go" Deliberations. Using this technique can reduce the adverse effect of cognitive biases, while controlling risk in dangerous operations.

To improve the quality of decision-making, some organizations use a similar technique known as "Red Teaming." A Red Team is formed to challenge the possible decisions independently. Both techniques are intended to influence higher-quality decisions in hazardous operations and improve the long-term performance of the organization.

Example: Launch Fever in the LCC

One of the wonderful characteristics of the launch controllers and managers operating in the Launch Control Center (LCC) at the Kennedy Space Center was their dedication to the mission of sending humans into space. Many of them devoted their entire careers to launching people on rockets. They didn't do it for the money; they did it to contribute to a noble cause.

With such passion to fly, being infected by a psychological malady we called "launch fever" may be human nature in the space business. I could hear it in the voices of the managers and see it in their faces when I worked in the LCC. At the moment the launch director was forced to scrub (cancel) a launch for the day, because of a technical issue or a weather problem, the members of the launch team let out an audible groan throughout the LCC.

As a flight crewmember who occasionally had the privilege of flying on the Space Shuttle, I was never comfortable hearing or seeing such an emotional release from the managers and engineers who were supposed to be making dispassionate decisions. High emotions have no place in the business of high-flying rockets.

Through my experience as a manager and commander, I learned to expect questions regarding emotions from members of the media after a launch scrub. The press corps always seemed to be sympathetic to what they perceived as the plight of the dejected astronaut who had to unstrap after a failed launch attempt. With bright lights, cameras, and microphones in the faces of the crew, they asked in a sympathetic tone, "Weren't you disappointed you didn't get to launch today?"

I was probably the most fortunate of all commanders. In five launches as a commander, I had to answer that question only one time after a single scrub when we had to unstrap. I don't remember what I said, but I know I gave a polite answer. In my head, though, I couldn't help thinking, *What a mindless question. Why would I want to launch on a broken rocket? I was elated they didn't make me launch. Now we can fix our problem.*

To me, the decision to scrub the launch for any reason should be viewed as a success, rather than a sign of failure. A problem was identified, options were discussed, attempts were made to rectify the problem within the rules and the timeline, and the correct decision was made to cancel the launch. The people did what they were paid to do, and the system worked. The crew lived to try it again another day.

The Space Shuttle Program worked best when George Abbey was running human spaceflight for NASA, which was after the *Challenger* accident and years before the *Columbia* accident. Mr. Abbey listened to anyone who had a technical or operational concern. When managers or engineers said, "Go," he asked many insightful questions to understand why they were confident. He had to be convinced we were really ready to launch. Mr. Abbey seemed to challenge the Go recommendations more than the No-go recommendations. He had a conservative philosophy, which was appropriate in our dangerous business. Yet, his questioning didn't slow us down. His conservatism made us better. He inspired operating excellence. We had the highest flight rate under his tenure, and every flight was successful.

24. Be Assertive (to Authority) When Necessary

Often during hazardous operations, managers must make decisions or take actions that change the level of risk. Wise managers solicit, encourage, and value the opinions of other managers or operators in the chain of command before making the decision. History has recorded many accidents that have occurred after a subordinate manager or operator correctly identified the signals of an impending accident but did not communicate effectively enough to the decision-making manager to prevent the accident.

This technique, *Be Assertive (to Authority) When Necessary*, is intended to encourage people to take action or speak to a manager or operator in a position of authority to improve operations. In some organizations executing this technique requires courage in speaking truth to power and standing up for convictions and principles. In better organizations courage is not required because managers in positions of authority encourage people to speak up. Good managers know the value of speaking up, and they show appreciation to people who do speak up.

The Occupational Safety and Health Administration (OSHA), under the US Department of Labor, administers a Voluntary Protection Program (VPP) to encourage organizations to provide safe and healthful working conditions. To be recognized as participants in the VPP, an organization's managers must demonstrate commitment by "Establishing lines of communication with employees and allowing for reasonable employee access to top management at the worksite."[31] To fulfill an additional requirement, "Employees must receive feedback on any suggestions, ideas, reports of hazards, etc. that they bring to management's attention."[32]

As a demonstration of compliance with these requirements, some organizations establish a process that allows personnel to speak up if they have a safety concern and request the work be stopped until the concern is addressed. Managers and personnel often refer to this process as "Stop-Work Authority" or "Stop-the-Job." On STS-113, we called this "All Stop."

Many airplanes have flown into the ground because subservient copilots were reluctant to question the decisions or actions of domineering senior pilots. Elite leaders welcome dissenting and challenging opinions because they know the added analyses will make the results stronger in the decision-making process. Create a Stop-Work Authority Program in your organization. Encourage your employees to execute stop-work authority when necessary. Give positive reinforcement for their actions when they do.

---

31 Directive Number: CSP 03-01-003, Voluntary Protection Programs (VPP): Policies and Procedures Manual (Washington, DC: US Department of Labor, Occupational Safety and Health Administration, 2008), 22.

32 Ibid., 23.

Example: All-Stop Process

To be used by all employees effectively, here is how a stop-work authority process should function and be supported by leaders.

The job site was in low Earth orbit. The Space Shuttle *Endeavour* was connected to the International Space Station. It was late in the mission, and three separate crews were conducting joint operations. Russian Cosmonaut Valery Korzun was the commander of Expedition-Five, the fifth crew in residence since the station became permanently inhabited in November 2000. American astronaut Ken Bowersox was the commander of Expedition-Six, which was relieving Five. Ken is shown in photo 126, floating in the *Destiny* laboratory module on the Station.

I was the commander of STS-113, with my crew on *Endeavour*. We were the taxi drivers, crane operators, or construction workers, depending upon which part of the crew exchange/assembly mission we were currently executing.

Photo 126: Commander Ken Bowersox, Expedition-Six (Photo Credit: NASA)

As an aside, having two active commanders on connected vehicles does not present a leadership dilemma. Nonoperating executives or managers might worry, but operating leaders know how to make the chain of command and decision-making authority work effectively, as illustrated in this example.

Toward the end of our joint mission, Ken was on the Station side and was controlling its arm. (The controls for the arm are shown on the far left side of photo 126. The Station's arm is shown in photo 127.) As Ken controlled the arm, flying it along a planned trajectory, I was observing from *Endeavour*. At one point, near the end of his planned sequence of motions, I thought the arm was a little too close to a fragile part of *Endeavour*. I picked up the mic and called over the intercom, "Stop, stop, stop." This is how Ken and I had agreed to initiate our all-stop process, which anyone could call for any reason.

Photo 127: Crane operations during EVA, STS-113 (Photo Credit: NASA)

Ken immediately stopped moving the arm. The first words out of his mouth over the intercom were, "Thank you very much for calling the all stop." Then he offered me a rather detailed explanation of his remaining trajectory, how much farther he would take the arm, what he was looking at to ensure safety, what systems he was monitoring, and how he was determining the exact distance to the fragile structure. In my opinion, he was exhibiting operating excellence at its finest.

At the end of his explanation, he offered me reassurance by saying two final things. First, he told me, "If you are satisfied with my explanation and you're confident that I won't damage the structure, and if you agree that I can continue safely, then I will, but I'll wait for you to give me clearance." He concluded by repeating his opening sentiment, "Thank you very much for calling the all stop."

That's how a Stop-Work Authority Program should work. Ken made me feel good about calling the stop, even though we concluded the work could continue. In Ken's mind, his satisfactory explanation did not prove I was wrong in calling the stop. Together, we proved his satisfactory explanation was needed before the work could continue.

Example: Autoland Discussion with Executives
Sometimes speaking truth to power can be extraordinarily difficult. Early in my NASA career, I was given the opportunity to experience just how challenging and uncomfortable this can be.

The Space Shuttle was designed with the capability to land automatically, in case the pilots couldn't see the runway. Managers in the Space Shuttle Program wanted to demonstrate this autoland capability during a Space Shuttle landing in good weather, before the system might be needed in bad weather.

The demonstration test was scheduled on STS-53 in 1992. In the year before the flight, the experienced pilots in the Astronaut Office evaluated the autoland system in the shuttle training aircraft, a Gulfstream II airplane modified to fly exactly like the Space Shuttle on approach to landing. During the training and evaluation flights and in the ground-based simulator, the pilots uncovered an interesting (otherwise known as dangerous) deficiency in the autoland system.

Most of the time, autoland functioned properly, with performance results usually better than those of the pilots. Occasionally, though, with certain navigation failures, the autoland performance indicated the Space Shuttle would have been destroyed with an excessively hard landing. In the shuttle training aircraft, by adding power to the engines, the evaluation pilots could disengage autoland and take over manually to prevent a hard landing. But the Space Shuttle's engines were not designed to function during landings, so a hard landing couldn't be avoided in that failure scenario.

Though the likelihood of experiencing navigation failure during a flight demonstration test on STS-53 was extremely remote, the pilots in the crew office concluded we should cancel the demonstration of autoland. Most of the time in the simulator, autoland performed better than the pilots. On rare occasions, autoland killed the pilots.

The program managers didn't buy our argument. They thought the failure scenario was too remote to worry about. The program managers, who wanted the demonstration, and the astronauts, who didn't, decided to elevate their disagreement to the center director, Johnson Space Center, for a final decision.

Though I was a rookie astronaut and had not yet landed the Space Shuttle as a commander, I was the crew office representative responsible for the landing and rollout systems on the Space Shuttle. My job, with little experience, was to plead our case and try to convince the senior-level board of highly experienced NASA executives we should cancel the autoland demonstration because there was a remote chance of causing structural failure of the Space Shuttle on landing.

On the appointed day, I put on my best suit and tie and walked over to the executive-level conference room. High-ranking officials from multiple NASA centers were in attendance. My technical presentation was about fifteen minutes long. I projected the data, analysis, conclusions, and recommendation on a screen, using several pages of viewgraphs (this was before we used PowerPoint).

About halfway through my presentation, immediately after I read one of my bullet points, the center director suddenly jumped up and started screaming at me. I sat stunned and at full military attention (if one can sit at attention), listened to him accuse me and every other pilot in the Astronaut Office of not being team players, and not supporting the program, and being unwilling to consider ideas for the greater good, and . . . I stopped listening after the first twenty seconds. I had a few minutes to think at light-speed, *What am I going to say when he finally sits down?* The only thought I had was, *I can't back down.*

The rationale for our argument was sound. This didn't mean I wasn't scared. I was pretty much scared to death, in front of a roomful of senior executives, with the most powerful one screaming at me.

Eventually, he did sit down, still seething, red faced, and angry. I tried to wring all emotion out of my voice. I referred to my charts, backed up two lines, and repeated my two previous bullet points that had sent him into a rage, as if to say, "I'm not changing my story."

I briefly looked over toward my boss, Mr. Abbey, and thought I saw him trying to hide a smile. I knew he supported me even if other senior executives didn't.

Soon after our meeting, the center director made the decision to cancel the autoland demonstration test on STS-53. The system was never needed.

---

*Example: Mission Management Team Round Table, Post-*Columbia *Accident*
I learned firsthand what going up against experienced managers to present unpopular technical arguments felt like to the junior engineers. And I had it easy because the crew office was not in the same chain of command as the program managers. I could say what needed to be said with low risk of career-limiting repercussions.

Through the next ten years, I saw many engineers preparing to make similar arguments, and they were physically shaking in fear in the moments before beginning their presentation. The power differential at the head table where decisions were made, with the senior executive at one end and the junior engineer at the other end, was extreme. To be in the room was to feel the tension. In the charged atmosphere, I never wanted to touch anything metal in that room for fear of fibrillating my heart. Some engineers didn't last long in their career before they left their dream job. We used to say the managers treated the young engineers as consumables, like the liquid oxygen and liquid hydrogen on the Space Shuttle, to be used until they are expended. The program will simply find more to replenish the tanks.

After the *Columbia* accident in 2003, the world outside NASA learned what the culture was like in the decision-making boardrooms. The *Columbia* Accident Investigation Board's report[33] described the pressure fairly well. After the accident, significant changes were made at NASA. The program managers brought in outside experts to assess the organizational culture and make recommendations for improvement.

Sociologists and organizational psychologists took one look at the decision-making process and made a quick recommendation. The power differential is too

---

33   Report, *Columbia* Accident Investigation Board, August 2003, (Washington, DC).

Photo 128: Newly designed round table for Mission Management Team, post-*Columbia* accident (Photo Credit: NASA)

great, they explained, and you're making it worse by using a long, rectangular table. You've got to reduce the emotional tension in the room and create a better environment for engineers to elevate their concerns. A circular table would be much more conducive to decreasing the tension, encouraging healthy debates, and creating a better decision-making process, resulting in more informed decisions.

NASA managers took that recommendation, and others, to heart. Seven months, and several hundred thousand dollars later, the program had redesigned the management team's meeting room. Better microphones were installed so the participants could hear the discussions clearly. Cameras were installed around the room in various places. The audiovisual system projectors and displays were improved, the better to see the PowerPoint presentations. And a large, circular table was installed as the centerpiece of the newly designed room.

Study photo 128. This is a picture showing the room from the point of view of the managers. Think about the shape of the table, and try to feel the reduced power differential the psychologists recommended.

Now look at photo 129. This is a picture of the room taken from the point of view of the junior engineer who is responsible for making an unpopular recommendation with insufficient data. Do you think you would feel psychologically safe to raise a concern?

The manager is responsible for encouraging people to Be Assertive (To Authority) When Necessary, and showing appreciation when they are.

The manager, not the table, is responsible for creating a better decision-making environment.

Photo 129: Round table, from the point of view of the presenter, often a subordinate engineer (Photo Credit: NASA)

Example: Landing Signal Officer Platform

We can look to the US Navy to find one of the best examples of using the technique Be Assertive (to Authority) When Necessary. The landing signal officer (LSO) stands on a platform on the port side of an aircraft carrier's landing area. The primary duty of the LSO is to observe the aircraft on final approach for landing and verify the pilot is flying within safe parameters. Photo 130 shows an A-7 Corsair approaching an arrested landing in a picture taken from the LSO platform. If the LSO determines the approaching aircraft is unsafe for any reason, the LSO illuminates the wave-off lights and transmits a verbal command over the radio: "Wave off, wave off, wave off!" The approaching pilot must immediately apply full power and go around for another attempt to land.

Photo 130: In the US Navy, the landing signal officer has the authority to wave off any airplane on final approach to an aircraft carrier.

The LSO may be the most junior ensign in the Air Wing. The pilot may be the most senior pilot, the commander of the Air Group. In that situation, it makes no difference. The junior officer issues the command, and the senior officer must comply.

This system of assertiveness works for two reasons. First, the LSO is highly trained. Second, the practice has become part of Navy culture. All naval aviators understand the value of having an LSO as the final authority on safety when the aircraft approaches the flight deck at 135 knots and the difference between a perfect landing and death is a little more than fourteen feet of altitude between the airplane and the ramp as the ship pitches up and down in the ocean.

Example: STS-52 Launch Decision

This final example is offered for you to consider what might happen in your organization when a stop-work request is denied. This is the story of the only launch in American spaceflight history that followed a No-go vote from the flight director in the Mission Control Center in Houston. Our example picks up on the launch pad of the STS-52 mission during our weather delay for high crosswinds. This is the second half of the story I began in the technique Balance Confidence with Humility, and it will end with the launch of STS-52 (shown in photo 108 and on the cover of this book). In the hour before that picture was taken, the launch managers and directors held contentious discussions about whether to launch or not. The decision to launch was complicated and challenging for those managers and directors.

The countdown clock was held at T minus nine minutes for about an hour. The peak crosswind at the nearby Shuttle Landing Facility runway was twenty-two knots, which was seven knots over the flight rule limit specified in the launch commit criteria. A landing would only be needed if we launched and then experienced an engine failure anytime before three minutes and forty-eight seconds after liftoff.

Coincidentally, in the year before this launch, I was the flight crew office representative responsible for evaluating the flight rule limit for crosswind. With extensive practice, I was able to land the orbiter in the simulation with twenty-seven knots of crosswind, even with the preconditions of blown tires on one strut and a nonfunctioning nose wheel steering system.

But, my ability to control a simulated landing with so much practice was not relevant for determining the flight rule limit for every launch of the Space Shuttle with all pilots in the Astronaut Office. When landing the real orbiter, any commander (including me on this day) will not have recently practiced hundreds of landings.

Based on the collective performance of all astronaut pilots with a normal amount of training, the managers, engineers, and astronauts made a joint decision to limit the crosswinds to fifteen knots. This was a conservative limit that would accommodate the skills of all the pilots and any unknown flight characteristics of the orbiter not modeled in the simulator.

On the launch pad, with the countdown clock holding, we settled in to wait for the decision. For most of the hold period, it was quiet on the radio and in the cockpit. After about an hour, the launch team began to transmit over the radio portions of the discussion regarding the crosswind. It gradually became clear to me that peak crosswind was still at twenty-two knots, but the manager from the Shuttle program, who was responsible for the decision, was considering rationale to launch anyway.

Essentially, he argued the flight rule limit should refer not to the peak crosswind value but, rather, to the integrated area under the curve of instantaneous wind speed as plotted over time. In other words, he asserted the orbiter would not respond adversely to momentary gusts of twenty-two knots but, rather, would be affected by the average wind, which was less than fifteen knots.

The flight director in Houston, who was responsible for the mission, disagreed strongly. He had participated in defining the flight rule limit, and he knew both peak wind and average wind were factored appropriately in the flying qualities evaluation that resulted in the decision to limit the peak crosswind to fifteen knots.

In the cockpit of *Columbia*, we could hear the deliberations. Near the end of the hold period, I began to realize the flight director was losing the argument, and the manager was going to commence the final launch countdown.

Several thoughts ran through my mind at high velocity. If the final decision was to launch, I was ready. Even though I had never landed a real orbiter before, I was confident based on my experience in the simulator I could handle the crosswind if we had an engine failure after liftoff and needed to return for an emergency landing. But what if the simulation did not model the true handling characteristics of the real orbiter? What if I lost control of the vehicle on landing?

How would managers make the decision? Would it be acceptable for them to launch us after overturning one year's worth of extensive and rigorous engineering analysis, which indicated the crosswind limit should be fifteen knots? And what would the managers do the next time with a similar decision?

I quickly concluded deciding to vote Go or No-go from this wider perspective of program managers, though important, wasn't in my job description. My responsibility was to decide if we were ready to launch from the limited perspective of the flight crew. I knew the managers and my flight director would be thinking about the higher-level programmatic considerations, and I decided I would be comfortable with whatever decision they made. I was ready to launch—or scrub—and conduct either sequence of operations flawlessly.

In the final polling the members of the launch team were asked to provide a verbal Go or No-go for launch. Each person issued a Go for launch until there were two members remaining in the poll: the flight director and me.

The flight director gave the manager his official answer for launch: "No-go." I was expecting this. And then he added the following amendment by saying essentially, "But if you're going to launch anyway, we are ready to support." I had not expected that. He had chosen his words carefully, and I was proud of him. I instantly knew he argued well but had lost.

After the flight I learned why the flight director made his decision. His main rationale to support the launch was the immense amount of time he and I had spent analyzing the flying qualities of the orbiter with crosswinds exceeding twenty knots. He knew the weight and center of gravity of the Space Shuttle *Columbia* on STS-52 were benign and not near the worst-case scenarios we evaluated. So he believed this operation would be safe on this day with this specific commander—even though I had never landed the orbiter before. But he also knew our analysis did not support raising the crosswind limit to more than fifteen knots on future flights. He knew the value of the rule, and he thought we shouldn't launch this day; still, he would support the operation with the best of his ability should the manager decide to launch us anyway.

Before the final poll commenced, the flight director had considered disconnecting his headset from the console and leaving the control room in protest to prevent the launch. In the end, he decided to remain and lead his team, confident that he and I would operate safely. Because the flight director didn't want the program manager to change the rule in the future, he voted "No-go." Yet, he knew the operation would be safe on this day. So he added his statement that, if the manager was going to overrule his vote anyway, he and his team were "ready to support" the launch. With those statements, I believe the flight director fulfilled his responsibility to my crew, NASA, and our nation to lead our mission safely, and I was ready to follow him.

As the commander of *Columbia*, I was the last to be polled. I had only a few seconds to think about my answer and its implications. Rather than using the standard words, "Go" or "No-go," I picked up on the precise words of my flight director, whom I trusted with my life. Over the radio, I said, "The decision is much harder for those of you who are supporting us; if you are ready to launch, *Columbia* is ready."

With the poll complete, the manager from the Program Office directed the launch team to resume the countdown. In the final nine minutes before main engine start, on three separate occasions, I thought to myself, *I can't believe they overrode the flight director!* Each time, I tried to force this extraneous and disconcerting thought out of my mind and focus on my job of commanding a Space Shuttle through ignition, liftoff, and ascent to orbit.

About two minutes before liftoff, I was relieved to realize that fourteen years of practice in hazardous flight operations had paid off, and the familiar feeling of directed focus of attention and heightened awareness of operations in the present had returned. When the picture on the cover of this book was taken, distracting thoughts about the management decision to waive the rule did not exist, and my entire universe was accelerating straight up at twice the force of gravity.

———

After-action reviews are rearward facing. Just as in incident investigations, retrospective analyses of management decisions to deny Stop-work requests are not appropriate for judging them as right or wrong. The participants always make decisions they believe are correct at the time. The relevant issue for us now is to understand why the decisions were deemed correct in the past and to look toward the future and decide what we will do when faced with a similar situation.

Here are the conclusions I have reached after thinking about this example for many years, not only as the commander responsible for my crew in the moment but also as a manager, who was later responsible for the entire cadre of astronauts in the program. Whether the operation can be conducted safely or not is only one aspect of the management decision.

If you are a manager who is responsible for the success of an operation, and you are considering denying a Stop-work (No-go) recommendation you have received, you should understand the implications of continuing the work from three different perspectives:

1. *The immediate, specific operational concern.* In viewing the issue from this first perspective, the basic question is simple though not necessarily easy to answer. As the manager who has received the Stop-work recommendation, you should consider the immediate and short-term implications for the operation if the work were to continue. The fundamental question is whether the work can and will be conducted safely or not.

    As you consider the specifics of the operational issue, you should endeavor to answer the following questions:

- What do I predict the short-term operational results will be, if I decide to continue the work? Will the operation be safe if I decide to continue the work?
- How do I know? Do I have accurate data? Am I considering all the data? Are my assumptions valid?
- Do I understand why the person or group has recommended the work be stopped?
- Do I believe the quality of my analysis made now is higher than the quality of the analysis made by members of my organization before now?
- Should I make allowances for the unique skills of the particular operators who will continue the work? What if an individual operator fails in the work?

No doubt there will be other considerations I haven't listed here. After viewing the issue through only this first perspective, you are not yet ready to make your decision. If you think the operation can continue safely, before you decide to deny the Stop-work recommendation, you should consider the issue using the next two wider perspectives.

2. *The effect on the assertive individual.* If you are the manager responsible for making the decision, you should consider the effect your decision will have on the individual who is exercising Stop-work authority.

If you deny that individual's request to stop work, you risk losing the contributions of that individual to your missions in the future in one of three ways:

- Depending upon the integrity and convictions of the individual, he or she may leave your organization.
- If the individual decides to remain in your organization, you may see that person abdicate his or her Stop-work authority in the future. Essentially, your decision has rendered that individual's Stop-work authority weightless.
- Depending upon how you execute your decision to deny the No-go recommendation, that person may not want to work for you or contribute to your success in the future.

If you value the contributions of the assertive individual, you need to weigh the potential effects on that person carefully.

3. *Long-term effect on the organization.* Finally, before you make your decision, you should consider the long-term effect you will create in your entire

organization. When the issue is assessed from this widest perspective, these considerations will present the greatest challenge for you as you make your decision.

You should ask yourself these questions:

- What will be the long-term effect on my organization if I deny the Stop-work request? How might my decision affect future contributions of members in my organization? Will my decision influence other members to accept greater risk in the future? Do I want them to accept greater risk?
- Will other valuable members in my organization abdicate their Stop-work authority—and other decision-making responsibilities—if I deny this request? Will I inhibit a healthy culture of speaking up? Will my organization drift closer to our next accident?
- If I deny the Stop-work request, do the short-term operational gains from continuing the work outweigh the possible long-term damage to the culture of operating excellence I'm trying to create and nurture? Do I really understand the long-term cultural implications of denying the Stop-work recommendation?
- Is this a one-time decision, or will I change the rules for making similar decisions in the future? Am I considering the capabilities of any of our operators who may be doing similar work in the future?
- On the other side of the argument, what will I gain if I accept the No-go recommendation, delay the operation, make some changes to satisfy concerns, and continue the operation later? How will personnel in future operations assess my leadership skills differently if I continue the work today, tomorrow, next week, or next month?

You may have to consider other important implications before making your decision. My advice is to consider these long-term questions and weigh their implications most heavily. These will be the most challenging. The payoff for valuing these long-term implications will be making the best long-term decision for your organization.

I have used this example of the launch decision made over the objection of the flight director because I hope the story will help you and your organization when you are faced with a similar issue. What really matters now is how you will decide when you are faced with an important Stop-work recommendation in your future. Your issue will be complicated and challenging. To help you make your decision, I recommend you consider the issue from the three perspectives of:

1. The immediate, specific operational concern;
2. The effect on the individual who is recommending Stop-work; and finally,
3. The long-term effect on your operations and the culture of your organization.

Every decision you make creates your future.

25. Be Cognizant of Limitations (in the Sociotechnical System)

Every sociotechnical system humans create has weaknesses and limitations on applicability or effectiveness in some elements of the system.

On the social/human side, these weaknesses and limitations may exist in the knowledge, skills, or attitudes of operators, managers, executives, teams, or the collective organization. Indications of the weaknesses and limitations are numerous and may include poor or biased decisions, inappropriate actions, ineffective communication, poor performance, insufficient awareness of the environment and operations, and lack of teamwork.

On the technical/systems side, these weaknesses and limitations may exist in the strategy, policies, rules, procedures, checklists, processes, workflow, activities, and the design or functionality of equipment, or others.

Astronauts use this technique, *Be Cognizant of Limitations (in the Sociotechnical System)*, to stay alive and be more productive. Before we launched on the STS-113 mission to the International Space Station, each crewmember knew his or her personal limitations, the limitations of the other crewmembers, limitations in the equipment, software, and systems, and the weaknesses and potential for errors in the procedures. In training, we developed a deep understanding of our personal and team limitations based on the mistakes we made. We understood the limitations of the technical systems and the procedures based on our performance in the integrated simulations with the flight controllers. From our understanding of the limitations in the sociotechnical system, we knew how to help one another during the mission and what techniques to use to eliminate nearly all mistakes during flight.

To control risk and improve productivity, operators must be aware of the weaknesses and limitations and take appropriate actions to mitigate the consequences of those weaknesses and limitations. The appropriate actions to take are various and depend on the specific weaknesses and limitations.

Previous Example: "Controlled Flight into Terrain" in the A-7 Corsair
The "Controlled Flight into Terrain" example I used in the technique Follow Procedures (and Rules) *Thoughtfully* is worth mentioning again.

Early in my high-hazard career, I began to realize the sociotechnical system, intended to help me control risk, contained weaknesses. Maybe my fear of death was causing me to be skeptical that the system would always keep me alive. I knew the system was merely a collection of humans, with inherent weaknesses, creating rules, with embedded weaknesses.

Though other pilots often complained our safety rules were overly conservative, I always assessed rules critically to understand if they would keep me alive. When I determined a rule was not conservative enough, I implemented self-imposed techniques or limits on the process, activities, or procedures being followed. This is what I chose to do after the US Navy imposed the new rule in the late 1970s following the A-7 Corsair accidents resulting in controlled flight into terrain.

To stay alive and be successful, operators in high-hazard businesses must Be Cognizant of Limitations (in the Sociotechnical System). You can see the technique Follow Procedures (and Rules) *Thoughtfully* is a specific application of this wider technique. Look for all potential weakness in the system. When you find weakness in the procedures, don't follow them blindly.

Example: Hazardous Driving Conditions

Before sunrise on December 2, 2014, I was driving into downtown Houston to deliver an address at a conference sponsored by OSHA. It was raining, the visibility was reduced, it was dark, and I was driving on unfamiliar roads (similar to the conditions

in photo 131). I instinctively reached up and turned off the radio to reduce potential distractions, and I ramped up my level of concentration on the driving task.

My increased emphasis on operating performance had nothing to do with my fear of being late to deliver a safety presentation at an OSHA conference because of a car accident, which certainly would have been an embarrassing story to share. Rather, my increased emphasis on the quality of my operating performance in a dangerous system had everything to do with my fear of dying and I am aware of the limitations of our traffic system on highways in American cities.

Photo 131: Low visibility when driving at night

To prevent accidents and improve performance, operators must Be Cognizant of Limitations in individuals, teams, policies, and equipment, and take appropriate actions to mitigate the adverse consequences of those limitations.

26. Assess Competence (in Team Members)

Successful operations require operators who are capable of performing their jobs. Leaders are responsible for ensuring operators are properly trained. To verify the training is effective, the leaders in the organization should develop and follow a process to *Assess Competence (in Team Members)*. This is critical in high-risk businesses. Competency assessment is conducted after building capability through training and practical experience, and it is the final part of the initial program of qualification to do a job.

Competency assessment is also the final part of a recertification process, which is the periodic element of a career-long program of building proficiency, or deep capability. In the assessment of competence (for the initial qualification and the periodic recertification), knowledge, skills, and attitude are evaluated. A primary source of evidence must be the performance of competency—the operator must demonstrate the necessary practical skills in the field, on the job, or in a simulator, and not in a classroom. These skills must be performed to specific standards under specific conditions to receive a certification to operate in the role, or a recertification to continue in the role.

Operators should continuously assess the competence of their teammates by using an informal process of assessment. When crewmembers correctly assess the competence of their teammates, they can more accurately capture and limit the errors made by the teammate and mitigate the consequences quickly. If a crewmember fails to understand the weaknesses of a teammate, or incorrectly assumes the teammate will perform error-free, that crewmember may miss opportunities to help the teammate prevent errors.

Example: Titov Speaks the Language of the Pilot
I had the privilege of flying in space twice with Colonel Vladimir Geogiyevich Titov, Russian Air Force (Ret.). As part of the joint partnership with the Russian Space Agency and NASA, Cosmonaut Titov was assigned as a member of my Space Shuttle crew on two missions, STS-63 and -86. To honor our commitment with the Russians, I didn't want Titov merely to ride along as a passenger. Rather, I wanted him to share important responsibilities on the crew and contribute to the success of the mission. So on STS-63 I assigned him the responsibility of deploying a Spartan satellite from the Space Shuttle *Discovery* (see photo 132).

Photo 132: Russian Cosmonaut Vladimir Titov flew as a mission specialist on STS-63 and was responsible for the deployment of the Spartan satellite (Photo Credit: NASA).

Some NASA managers were concerned with Titov's limited command of the English language. I wasn't. As the commander, I was responsible for assessing competence. The first time I flew with Titov in the simulator, I was convinced he understood operating excellence as a pilot even if he didn't understand much of the English language.

At one point during a simulation, we were faced with a challenging emergency deorbit scenario. This required each crewmember to work independently and expeditiously to configure different systems on the orbiter with many switches in the cockpit. Titov was using a checklist written in English, which he may not have understood well. But his performance was flawless. At one point he demonstrated masterful technique when it was time for him to configure a critical system, which would result in loss of the vehicle had he made a mistake. Appropriately, he waited until another crewmember could assist him to ensure error-free operations. His intuitions, based on his experience as a pilot, were perfect. Immediately, I realized Titov spoke the "language of the pilot" in our dangerous business. When mistakes can kill, being fluent in *pilot* is better than English or Russian.

About five months before flight, a training event was scheduled in the simulator to allow Titov and our crew to practice deploying the Spartan satellite. As part of our agreement with the Russians, operations on the Space Shuttle would be conducted in the English language using checklists written in English. Operations on the Soyuz would be conducted in Russian using checklists written in Russian. That afternoon our crew went through the Spartan deploy procedures six times. Titov failed to perform the procedures correctly on five of those six times.

By this point, NASA managers were really worried and approached my boss with their concerns about Titov's language capabilities as a Space Shuttle crewmember. Throughout training, I continued to assess competence. I remained convinced Titov was highly competent. The failure, in this case, was ours. I told the managers, "Two things are going to happen. First, Titov is a pilot, and he does not want to make these mistakes again, so he will study. Second, you will allow us to train him using the Russian language until he understands the Spartan system. In one month, we will repeat the training event, using English, and he will be successful."

One month later, we repeated the training exercise. Titov performed correctly on twenty-five out of twenty-five runs in the simulator.

On both space missions, Titov performed flawlessly and contributed greatly to the success of the missions. On STS-86, our second flight together (see photo 133), Titov became the first Russian cosmonaut to perform a spacewalk in an American suit from the Space Shuttle.

I take no credit for Titov's superior performance. He came to us as a great operator. I will take credit for properly using the technique Assess Competence (in Team Members). My assessment resulted in giving him greater responsibilities, against the recommendation of some NASA managers. Titov's individual performance and teamwork improved the collective operating performance of our crews on the two missions.

Photo 133: Russian Cosmonaut Vladimir Titov, STS-86 (Photo Credit: NASA).

27. Acknowledge (Personal) Weaknesses

Groups of people working together can accomplish much more than individuals working alone. Even with the same numbers of people, a group working together can outperform individuals working separately because the group can overcome the weaknesses of some members with the strengths of other members. Of course, this enhanced performance of the group depends on knowing the members' weaknesses.

Operators who *Acknowledge (Personal) Weaknesses* to the team are helping the team—and themselves—perform better. The leader of the team can divide and assign the responsibilities effectively among individual operators to maximize strengths and minimize weaknesses to improve overall team performance. Operators with certain strengths can provide assistance and use their skills to help individual operators with particular weaknesses to reduce errors and improve the performance of the individuals, as well as the team.

Though this technique, Acknowledge (Personal) Weaknesses, is one of the simplest for people to understand, it seems to be one of the most challenging for people to enact. The reluctance to admit weaknesses begins in childhood and becomes stronger when interacting with peers in school. After leaving school, most young adults enter the workforce where good performance is rewarded. Poor performance is not. Too often in organizations, acknowledging weaknesses seems to be equated with admitting poor performance. The higher an employee wants to rise in the organization, where winning and succeeding are valued, the more motivated the employee may become to show only strengths and hide weaknesses. To the employee, there seems to be little advantage in admitting deficiencies or shortcomings.

The willingness to focus on flaws is much greater in operators who are part of elite teams engaged in dangerous activities. In high-performing teams conducting high-hazard missions, acknowledging weaknesses is necessary to survive. This is another reason I like working with operators who are working with risk. Life-and-death situations seem to bring out honesty in people.

But even for operators in hazardous operations, acknowledging weaknesses is not easy. When I became director, Flight Crew Operations, for human spaceflight at NASA, I realized some astronauts were beginning to behave differently when I flew with them in the T-38 aircraft. As the director, I was responsible for recommending astronauts for spaceflight assignment. I perceived some astronauts became reluctant to admit making mistakes or to acknowledge weaknesses to me because of their assumption that if they showed weaknesses I would not recommend them for assignment to a space mission.

That assumption was incorrect. In fact, the reverse was true. When I flew with astronauts who readily acknowledged they were working on a particular weakness, their personal stock went way up in my opinion. We all have weaknesses. I valued astronauts who understood and acknowledged personal weaknesses, knew the limits of their capabilities, were eager to learn, and wanted to develop methods to mitigate their weaknesses.

Early in my career I learned leaders should set a good example by acknowledging their own weaknesses to the team. But after I became a leader, I realized demonstrating the right behaviors by admitting weaknesses to operators did not motivate those operators to acknowledge their weaknesses to the leader. Admitting weaknesses to the team is far easier for a leader whose reputation has already been established. Operators are still trying to build their reputations, so admitting weaknesses is much more difficult. The leader must do more than solicit and encourage the acknowledgment of weaknesses. The leader must create a healthy supportive environment in which the operators feel rewarded and that their reputations are enhanced when they acknowledge weaknesses.

When individual operators understand the value of this technique, the team can use the technique collectively to reduce errors, prevent accidents, improve performance, complete dangerous missions, and stay alive to attempt future missions.

Example: Tasking for MS-2 on STS-113
On my sixth mission, I was looking forward to flying with John Herrington. He was on his first mission. I always enjoyed seeing the first-time flyers experience the wonders of spaceflight. John was our mission specialist-2, which meant he was our flight engineer and navigator. In addition to these primary duties, the MS-2 was responsible for providing assistance to the pilot and commander during the critical phases of flight (ascent, rendezvous, docking, reentry, and landing). John is shown on the left side of photo 134, in his role prior to our deorbit burn for reentry on STS-113.

When John was initially assigned to my crew, he asked me how he could assist in the best way. With that single question, I knew he would be a valuable asset on our team. I told him he had one important job. He was to understand my weaknesses and help me avoid errors. Since that was a huge responsibility, as I have many weaknesses, I told him I would try to make his job a little easier by acknowledging my weaknesses.

Photo 134: Mission Specialist John Herrington assisting Commander Jim Wetherbee and Pilot Paul Lockhart during deorbit prep procedures on mission STS-113 (Photo Credit: NASA)

But, I still wanted him to observe my performance critically and try to identify weaknesses I had not recognized or acknowledged.

John had a great flight and performed three outstanding spacewalks on our mission as we continued the construction of the International Space Station. And, he helped me mitigate my weaknesses.

28. Admit Errors

Committing errors is a condition of being human. In hazardous endeavors, successful leaders and operators share the common characteristic of understanding they can and do make errors. In this context, errors are the commission of inappropriate actions, the omission of appropriate actions, or mistakes in cognition, judgment, or decision-making.

The best operators and the highest performing teams use specific techniques to limit the number and severity of the errors they make. Even with these techniques, great operators rarely perform error-free in complicated operations. But these elite operators almost always eliminate the consequences of errors completely or reduce the consequences to near zero, especially when it matters most in dangerous environments.

The best leaders and operators believe in the importance of this technique, *Admit Errors*. When an error is admitted quickly and candidly to teammates as soon as the error is recognized, corrective actions can be taken to mitigate the consequences more effectively and completely.

I think too many executives, managers, supervisors, and operators are reluctant to admit errors. Some even take active steps to hide errors they have made. In dangerous operations, hiding errors is unconscionable. But for the personnel trying to succeed within the cultures that have evolved in many organizations, hiding errors is also understandable and predictable. Too few executives and managers seem to understand the value of admitting errors in businesses that operate with hazards. Even fewer executives and managers know how to embed mechanisms to reinforce the desirable behavior of admitting errors.

Leaders in the operating organization are responsible for creating an environment in which all personnel understand the value of admitting errors and anticipate a positive experience when they admit errors because they are contributing to the success of the mission.

Example: Relationship with Instructors and Flight Controllers
In the second half of the preflight training syllabus before Space Shuttle missions, each crew participated in several integrated simulations with the flight controllers in the Mission Control Center. In these advanced training sessions, the flight crew and

the flight controllers operated as an integrated team. Together, they tried to identify failures and solve emergency situations inserted into the simulation by instructors, who were quite good at playing their devious role of trying to kill us (and, of course, making us stronger).

In my manager's office, I had a communications receiver that allowed me to monitor the debriefing sessions after each simulation. For several years, I heard how other crews interfaced with the flight control teams through their conversations. Sometimes I was disappointed to hear crewmembers blame the flight control team for failing to provide adequate help as the crew battled the emergencies during the simulated flight. Occasionally a crew made an error and blamed the flight controllers for failing to help them prevent their own error.

On the later flights I commanded, I enforced a no-blame policy during the debriefing conversations. None of us on the flight crew were allowed to blame a flight controller. Even if the flight control team made an error, I wanted us to look inward and analyze how we failed to help the team avoid their error. We acknowledged their good performance, and we admitted all of our errors.

With our no-blame policy and showing the flight controllers appreciation for helping us, I was merely trying to avoid the embarrassment of looking unprofessional. I soon realized there was a huge upside I had not anticipated. The flight controllers began to respond in kind. They had always admitted their errors. But with us, they seemed to emphasize how they failed to help us, even when the error was ours. I had the sense they really enjoyed working with us. They seemed to spend much more time and effort trying to help us succeed. Our working relationship, built on mutual respect, was out of this world.

By the time I flew on my sixth mission, thanks to the tremendous support we received from our instructors and the flight control team (see photo 135), we

Photo 135: Some of the out-of-this-world instructors and flight controllers for STS-113 (Photo Credit: NASA)

were the best-trained crew I ever saw. On STS-113, with exceptional help from our flight control team, we made the fewest errors of any flight I had ever flown or supported. We also accomplished more than many other missions. STS-113 was the first mission to the Space Station with double primary mission objectives of crew exchange and construction. I will always believe we were so successful because our instructors and the flight control team gave us such wonderful support in preflight training and during the mission. Maybe they enjoyed working with a crew that always tried to admit errors.

29. Use Methods to Aid Weak Prospective Memory

One shortcoming in our human brains is weak prospective memory, the ability to remember necessary actions in the future with no intrusive cues, alarms, or reminders. I deal with this weakness nearly every day of my life. And it's not only ordinary or monotonous actions that can be forgotten easily. Pilots can forget to lower the landing gear prior to touchdown on the runway. Before closing the body of a patient, doctors can forget to remove a surgical tool. Tragically, loving parents continue to demonstrate weak prospective memory when they leave their infant in a hot car on a summer day. None of these are criminal activities, and none of these reflect poor character. It is simply the way our brains work.

Fortunately, our brains have strengths in other ways, and we are able to compensate for having weak prospective memory. As operators, we can create reminders. Depending upon the operational situation, with a little ingenuity, we can create nearly infallible reminders.

On STS-113, we used simple methods to combat distractions and mitigate the consequences of weak prospective memory. Before important and time-critical events, we reminded each other of the impending event on regular intervals. If a maneuver or a cool-down procedure were to start thirty minutes in the future, we would announce the *time-to-go* each minute before the event. By enacting this technique, *Use Methods to Aid Weak Prospective Memory*, we never missed a time-critical event, even with all the operational distractions we were experiencing.

Example: Wearing Multiple Alarms

In spaceflight, time is limited. What goes up must come down. Actually, in orbit we are always falling and running out of time. With so much to do and so little time, we cram as many tasks as we can into our operational timeline. This means we are often conducting multiple operations simultaneously.

Unfortunately, this also means we may have multiple important actions to take for different operations with no intrusive cues to remind us when to take particular actions.

Said simply, forgetting to do something important is easy—even, incredibly, when our lives depend upon remembering. To solve this problem in space, I often wore two or three watches with multiple alarms (see photo 136). Sometimes, I had four different alarms set to remind me to take particular actions. The only problem I had then was trying to remember which action was related to which alarm—more difficult than it seems. (For critical actions, writing a note on my arm with a marker helped.)

Photo 136: Wearing devices (multiple watches) to aid weak prospective memory, STS-113 (Photo Credit: NASA)

Example: Gear Down

In aviation, many actions can be forgotten easily, which could result in accidents. One obvious example is forgetting to lower the landing gear prior to landing. If you are a pilot and don't think it can happen to you, someday it *will* happen to you. It is incredibly easy to be distracted momentarily and ruin your career in an instant (along with the underside of the airframe).

The problem was worse in the Space Shuttle. Because the orbiter had high aerodynamic drag and no thrust to compensate, we couldn't afford to lower the landing gear, which added more drag, until a short time before touchdown. As a normal operating practice, the pilot lowered landing gear at an altitude of three hundred feet above the ground, a mere ten seconds before landing. If a malfunction in the pilot's system prevented the gear from deploying, the commander could lower the wheels with independent controls. As is typical for well-designed NASA systems, we had redundancy to overcome failures.

What I worried about was a simpler failure. What if the pilot forgot to lower the gear? The ultimate solution for either the system failure or the human error was the same. The commander would lower the landing gear, but there was very little time to take corrective action. While rushing to look down and find the landing gear pushbuttons, the commander who was controlling the hard-to-control vehicle could easily land hard and damage the structure.

As a commander, my solution was to develop the habit of always deploying the landing gear redundantly from my side on every practice approach in the shuttle training aircraft (as shown in photo 137) and on all five real approaches on the Space Shuttle. I solved three problems simultaneously. If we experienced a system failure on the pilot's side, I had already deployed the gear from my side. If the pilot forgot to deploy the gear because of weak prospective memory, again, I had already deployed the gear. And finally, since I had practiced deploying the gear on more than one thousand approaches in the Shuttle training aircraft, I was no longer worried about landing hard while scrambling to find the gear pushbuttons after the failure.

Photo 137: One of more than a thousand practice approaches to landing commanders receive in the Shuttle training aircraft (an extensively modified Gulfstream-II) (Photo Credit: NASA)

Example: An Ingenious Reminder to Retrieve a Lunch Bag

Here's an everyday example. Astronauts must maintain physical fitness in preparation for spaceflight. One of my colleagues preferred to work out in the morning before arriving at the office. Rather than leaving his lunch in a hot car while he exercised, he stored his lunch temporarily in the refrigerator we had in our workout facility.

On one particular day, when his physical training was complete, he drove to work for his scheduled activity of learning how to not die in the simulator—always a fun activity. It wasn't until lunchtime he realized he forgot to retrieve his lunch from the refrigerator at the gym. From then on, he deposited his car keys in his lunch bag before he stored the bag in the refrigerator. I thought that was an ingenious method to aid weak prospective memory.

Example: Terminating Nonstandard Fuel Transfer in the T-38

When flying in the T-38, we occasionally experienced a spontaneous imbalance of fuel in our tanks, which shifted the center of gravity of the aircraft and created a potentially hazardous situation. Our solution was to isolate the lighter tank and burn fuel only from the heavier tank until the levels were equal. This configuration of the valves that control fuel flow set us up for a more dangerous situation. If we forgot to reconfigure the fuel valves when the levels were balanced, we would soon create an opposite and worse imbalance than we had in the beginning.

I knew someday I would forget to terminate the nonstandard transfer of fuel and create a worse hazard. My method to aid weak prospective memory was to remove a flight glove from one hand while I had the valves in the nonstandard configuration. This always made me a little uncomfortable to be flying without proper protective equipment (my fire-resistant glove). But I never forgot to reconfigure the fuel valves. I decided to accept slightly increased risk in one area to reduce the greater risk in another, and I figured a little psychological discomfort was helpful. I never wanted to be completely at ease in my dangerous environment.

30. Demand Operating Excellence from Myself First (Then Inspire Others)

Controlling risk and improving performance in our modern dangerous world with ambitious objectives—under dynamic, complex, and uncertain conditions—requires operating excellence from all personnel contributing to the mission.

As individual operators, none of us has the ability to influence the entire organization to exhibit the proper behaviors while following the highest standards of excellence in operations. Even leaders, who have wider circles of influence, may not be able to influence every member of the organization to strive for or practice excellence in operations.

Every one of us has complete influence over only one person in the organization: our self—and that's where we start. Each of us must *Demand Operating Excellence from Myself First*. Each of us has a much greater ability to influence the success of the mission if we personally use all thirty of these *Techniques for Operating Excellence* to prevent accidents and improve operating performance in ourselves first.

Leaders have a responsibility to create the conditions necessary for every member of the organization to strive for excellence and high-quality operations. Executives and managers must use their leadership skills to embed organizational mechanisms that reinforce the values and behaviors of operating excellence. When the members of the organization share the values of operating excellence collectively, and believe enacting those values will make them successful, a culture of operating excellence is in place. The culture becomes a force multiplier for the leaders. Even when the leaders are absent, the culture can influence the managers, engineers, and operators to exhibit the behaviors and practices of operating excellence.

Ultimately, none of us can inspire anyone else unless we first believe in and demonstrate through our behaviors the importance of operating with the highest standards of excellence. This is especially true for leaders. If members of the workforce observe the behaviors of any leader who doesn't understand or deeply value operating excellence, the desired culture will be destroyed quickly.

Everyone in the organization can contribute to improving and nurturing the culture. There is one certain way to help. In dangerous operations, each of us must Demand Operating Excellence From Myself First (Then Inspire Others).

Example: Supervisor at Plumb Brook Station, Near Sandusky, Ohio

In 2006, I visited a large vacuum chamber (shown in photo 138) at the Plum Brook Station, a NASA facility near Sandusky, Ohio. Historically, this chamber was used to conduct thermal-vacuum testing of the *Apollo* spacecraft when our nation explored beyond Earth orbit. During my visit, I had a wonderful opportunity to meet the supervisor, who led the maintenance team that cared for the chamber. My conversations with the supervisor led me to believe he was an outstanding leader.

When I meet people with exceptional skills, I try to learn from them. This supervisor and his crew had a reputation for performing their maintenance work with high quality. They caused few incidents or issues as they kept the facility in a high state of readiness. To understand his leadership style, I asked the supervisor this question: "In the rare instances when your workforce had an incident with

Photo 138: Thermal-vac testing of Apollo spacecraft was conducted in this giant vacuum chamber at Plum Brook Station near Sandusky, Ohio.

the equipment, or sustained an injury, what was the cause of that problem?" The supervisor smiled. I learned long ago when a leader smiles after a question, he or she is about to say something profound. I quickly picked up my pen and notebook to capture this particular leader's wisdom. He shared the following thought: "Problems in my workforce occur for one of only three possible reasons: I didn't explain the job requirements well enough, or I didn't give them proper tools or adequate time to complete the tasks, or I put them in a situation they weren't prepared for."

I suddenly realized, in each of these three self-aware conclusions, this superb supervisor had the humility to look in the mirror to identify the problem. He didn't blame his crew for mistakes or failures. He truly believed he caused the problem and he failed to help his crew succeed. His crew loved working for him. They readily admitted any errors they made because they knew he would protect them. This meant they learned more quickly, which resulted in even fewer errors. They asked for training when they needed it, knowing he would find a way for them to receive training. They readily asked for more time or better tools to do their jobs.

The upside of his leadership behavior was his maintenance crew loved working for him, they worked smarter and better, and they accomplished more work with fewer issues. To help me remember his leadership philosophy, I wrote in my notes, "Requirements, Resources, Readiness." I will always remember the three *R*s from this inspiring supervisor, who exemplified the technique of demanding operating excellence from himself first. He believed, *I set inadequate Requirements, I gave insufficient Resources, I exceeded their Readiness.*

Example: Welding Supervisor on an Offshore Oil Platform
Over the years I have realized some of my best conversations with front-line personnel in operations have occurred at 2:00 or 3:00 a.m., when the middle-level managers are usually sleeping.

In 2007, on a beautiful starlit night with a warm Caribbean breeze, I happened upon one of the best team leaders I ever met, Mr. Riaz Khan. He was the supervisor of a five-person contractor team responsible for performing welding on the oil-and-gas platforms in the Caribbean during the turnaround maintenance periods, similar to the work being conducted in photo 139.

Instantly, I could tell Riaz was a true student of human behavior. Our conversation afforded me a great opportunity to learn from him. I began to understand why his crew was so successful and why they had such a good reputation for high performance.

Photo 139: Welding crew on an offshore oil platform in the Caribbean Sea

Riaz explained to me, "If my crew welded ninety-five feet yesterday, I don't ask them to please weld one hundred feet today. I ask them to weld correctly. I tell them I will help them learn how to weld correctly." He never gave them a numerical target for performance. He only gave them a quality goal.

I asked Riaz how his team members were doing. He smiled broadly, in his Trini way, and proudly told me they were doing exceptionally well. They were becoming expert welders. Because they made fewer mistakes, they had fewer failed inspections so they rarely needed to repeat work. The upside was almost every day they did exceed the length they completed the day before. If middle-level managers levied performance targets measured by quantity, Riaz wisely insulated his team and refused to pass down quantity targets to his front-line workers. Through inspiring his workers to achieve higher quality, his team was able to achieve the targets of greater quantity the managers often demanded.

———

Demonstrate a consistent desire to improve. Be willing to receive and give help. By following these tenets of operating excellence, our crew minimized errors during our STS-113 mission in low Earth orbit. Many times during flight the individual crewmembers could have acted alone, without assistance from the ground teams or other crewmembers. But because we honored the simple desire to have help with procedures, no matter how small or unimportant, we were able to minimize our errors. Though having a desire to receive help does not sound profound, when all crewmembers displayed this genuine desire, the beneficial results were significant. Nearly all errors were prevented during the complex mission over two weeks.

Example: Paul's Integrity, Selflessness, and Desire to Do It Right

As my final example in this list of thirty *Techniques for Operating Excellence*, this is a short story about my copilot on STS-113, Colonel Paul Lockhart, US Air Force. Paul is shown with me in photo 140 on the flight deck of *Endeavour*.

Even though I had five previous space flights in my logbook and he had only one, I feel I learned much more from him about operating excellence, integrity, selflessness, and the desire to do it right than he did from me.

Photo 140: Pilot Paul Lockhart, with the author on orbit, STS-113 (Photo Credit: NASA)

By the end of 2002 when I flew with Paul, Americans had been flying in space for more than forty years. The techniques astronauts used to create a culture of operating excellence were well established, though not written down comprehensively. There is a long list of astronauts who understood and exemplified operating excellence. I wish I could have had the privilege of flying with each of them. I know I would have learned much from them.

Here is what I learned from Paul. We had just arrived on orbit. We were conducting our postinsertion procedures to configure the systems on the Space Shuttle *Endeavour* for orbital flight operations. I noticed immediately Paul was interested only in the success of the team and had no desire for any personal recognition. He exemplified the characteristic of selflessness. For the next thirteen days, he was only interested in how he could help the team succeed in our mission. With all my flight experience, I found myself trying to emulate the attitude of an astronaut who was on his second mission in space.

I began this book by writing about how important mental attitude is on the journey toward operating excellence. Of course, knowledge and skill are important in hazardous operations. But, knowledge and skill are defined, delivered, practiced, and emphasized in many aspects of training and operating in many organizations.

I have long felt attitude, the most important of the KSAs, was insufficiently considered in training and operations in many organizations. This is why I believe these thirty *Techniques for Operating Excellence* are so important. Each is about having the appropriate attitude to achieve operating excellence.

Of the thirty, this final technique may be the most important: Demand Operating Excellence from Myself First (then Inspire Others). That's what I watched Paul do for thirteen days in space.

And finally, on December 7, 2002, it was time for me to return to Earth for the last time. Photo 141 shows a long-range video capture of *Endeavour* with her crew, minutes before completing the mission of STS-113. We used all the techniques, which helped us control risk and improve our performance.

The practice of operating excellence, the basis for these techniques, was taught to me directly and indirectly by every astronaut who flew before me, beginning with the *Mercury 7* astronauts and including the newer astronauts who had not yet flown.

More than working with techniques, though, missions succeed because people work together. And it's not only the operators facing the hazards who create success. Support personnel behind the scenes contribute to the success of the mission by demonstrating operating excellence. Their job is more difficult. Climbing on top of a rocket is easy for the flight crew. We live or die as a consequence of our own decisions and actions. When the support personnel do their jobs, they must live with the consequences of their decisions and actions that cause other people to live or die. They deserve tremendous admiration for being able to perform under that kind of pressure and a big share of the credit for the success of the missions.

Photo 141: Long-range video of *Endeavour*, STS-113 (Photo Credit: NASA video capture)

Hundreds of thousands of NASA employees and contractors put their hearts and souls into America's space program. I thought about them the first time I sat on the launch pad as the commander of a space mission. Their dedication gave me the confidence to fly.

In the end, missions succeed because people work together. If you want to learn how to control risk and accomplish missions, study the people who work in dangerous environments. They are committed to the mission. They care deeply about the people who are contributing to the mission. They are attuned more to their external environment and the needs of others than their internal personal concerns. They learn quickly and they share widely. Those are the people who have made my career in dangerous occupations so rewarding.

There are some people who say the only thing that matters is results—achieving goals and accomplishing missions. I understand the emotional sentiment and the rational thinking behind this point. I certainly derived professional satisfaction from completing successful missions. But I have to say, I remember my experiences working with people much more than completing the missions. I valued the process of training together, learning together, and operating together more than the postflight celebrations of mission success.

After more than half a century, I have confirmed what I anticipated when I was ten years old. I loved being an astronaut. I still feel great satisfaction from operating successfully in dangerous endeavors, not because I like risk, but because I love controlling risk. I have had the best jobs—on and off the planet.

And the wonderful part is, I continue to have opportunities to work with amazing people who are passionate about controlling risk, improving performance, accomplishing great missions, and making our dangerous world a little better for the people who will come after us.

ABOUT THE AUTHOR

Jim Wetherbee

- Naval Aviator; Captain, US Navy (Ret.)
- Former Astronaut; NASA
- Former VP, Operating Leadership; BP Corporation North America, Inc.

With thirty-five years of experience in high-hazard operational environments, Jim offers guidance to leaders and operators in dangerous endeavors with critical mission objectives. He is the only American astronaut to have commanded five missions in space, and the only person to have landed the Space Shuttle five times.

Jim earned a Bachelor of Science degree in Aerospace Engineering from the University of Notre Dame in 1974. He began his career as a Naval Aviator aboard the USS *John F Kennedy*. Jim flew two tours in the Mediterranean flying the A-7 Corsair in VA-72, a light attack squadron. After graduating in 1980 from the US Naval Test Pilot School in Patuxent River, MD, Jim performed flight testing of the F/A-18 Hornet for three years.

In 1984, Jim was selected to join NASA in its tenth group of astronauts. Over a twenty-year career, he flew six times on the Space Shuttle. The five-time commander flew two missions to the Russian Space Station, *Mir*, and two missions to the International Space Station. In 1998, he was appointed as director, Flight Crew Operations, specifically selected to improve the flight and ground safety in the astronaut corps. Based on that success, Jim was selected after the *Columbia* accident to enhance the safety aspects in the organizational culture at the Johnson Space Center, home of NASA's human space flight program.

Bringing his experience from the aerospace industry as a former NASA executive and astronaut, Jim joined the oil-and-gas industry in December 2006 as a Safety and Operations Auditor for BP. He served two one-year tours as a senior leader helping with operational leadership and culture at a refinery and a drilling business. Later, as the VP, Operating Leadership, Jim supported efforts to improve performance results consistently over the long-term by emphasizing leadership behaviors to inspire people to conduct safe, compliant and high-quality operations.

After successful careers on the ground, at sea, and in space, Jim is passionate about helping leaders and operators perform successfully in hazardous environments.

Organizations:
- Lifetime Member of the Society of Experimental Test Pilots (from December 1983)
- Honorary Member, Musicians' Union, Local 47, American Federation of Musicians, Los Angeles, CA.
- US Astronaut Hall of Fame (inducted 2010)

Honors: Distinguished Flying Cross; Navy Achievement Medal; two Meritorious Unit Commendations; six Space Flight Medals; two Outstanding Leadership Medals; four Distinguished Service Medals.

ACKNOWLEDGMENTS

The list is long. I owe so much to the people who have taught me about risk.

I start with the two people in the world I admired most and who instilled in me a wonderful attitude for controlling risk, my parents. Dan was a nineteen-year-old pilot-in-command in B-25s on twenty-five missions over Germany in World War II—and then had the courage to sign up for a second tour in B-26s. After forty-eight missions, they told him he had risked enough, and they sent him home. I'm glad they did, or I probably wouldn't be here to write this sentence. As the director, Flight Operations, for the eastern region of American Airlines, he gave me my love for aviation. From Pop, I learned what it means to be professional in everything I do. There is really only one way to do things. When I had tough decisions to make in my leadership roles, I only had to think about how he would have accomplished the job or the mission.

Althea was an Army nurse in WWII, seeing action near the front lines in France and Belgium. Mom gave me my love for music. She took piano lessons for seventy years of her life but not to play in public; she just loved to play. Mom was an athlete before it was cool for women. Later in her life, when it was, she was the coolest. In the 1980s, she held nine national records for distance running in her age group. She won the New York marathon four times in the sixty-plus age group and added London, Tucson, Washington, New Orleans, and Houston for good measure. Mom saw only the good in people she met. She was incredibly optimistic, and she was funny—even in the last week of her wonderful life.

When I was seventeen, I knew if I waited long enough, I would find the girl of my dreams. At twenty-six I met Robin, and it was love at first, second, third, etc., sight. When she married me before I applied to NASA, she didn't know what she was getting into. Robin had the toughest job on the planet, watching a loved one climb on top of a giant stack of explosive energy and launch into space. Six times. I couldn't do it once if the roles were reversed. Robin has a personality that draws people in. I have seen countless strangers on the street single her out in a crowd and ask for help or directions or advice or just to have a conversation. She keeps me grounded. After one mission in space, when the final guest had left our landing party, she looked at me and said, "Great flight, big boy; now take out the garbage." Gotta love her. And I do.

My parents-in-law, Harry and Dot, loved me and I loved them. They had a wonderful life together.

I only wanted girls, and we were blessed with two. I wish I had Kelly's strength and tenacity. She knows what she wants and goes after it. I never worried when she went off to war with the US Army in Iraq—at least, not for Kelly or our side. I admire her business savvy and graphic design skills. She must have inherited that from her mom. I wish I had Jennie's empathy and pure intelligence. As a doctor of audiology, she has already helped more people in her short life than I have in my long one. I have the same unspoken connection with her as I had with my father. And what we do speak about is pretty awesome, too. I couldn't have asked for better sons-in-law with Umar and Kevin.

For sharing their lives with me and making me a better person, I thank my brothers, Larry, the golf pro, and Danny, the Hollywood film editor, and my sister, Dr. Jule, the cardiologist, and their spouses, Carolle, Annie, and Dr. Mike, and my brother-in-law and spouse, Tommy and Carol.

Mrs. Johnson and Mrs. Minot, my fourth and fifth grade teachers didn't chastise me when I smuggled a nine-volt transistor radio into class to listen to the *Mercury* astronauts reporting from Earth orbit. Rather, they allowed me to be the class rep to the space program. How great is that?

Steve Andersen gave me my love for risk control and rock music.

All my teachers at St. Hugh of Lincoln Elementary, Holy Family Diocesan High School, and the University of Notre Dame du Lac nurtured my love for math, science, and engineering.

I tried to learn the business aspect of risk from Doug, Tom, Joe, Stoltzie, Ed, Finn, Rico, JZ, and all others in the national championship class, ND '74, plus Hoff.

Frank Lizzo taught me to use the other side of my brain on the drumset and take risk on stage.

Prof. Marcelo R. M. Crespo da Silva, at the University of Cincinnati, gave me the graduate-level theories, knowledge, and analytical applications to understand the dynamics and control of space vehicles. Your theories and analysis worked in orbit, Doc!

Capt. H. A. (Tony) Merrill, US Navy (Ret.), taught me to be a naval aviator and military officer. I couldn't have asked for more, but then he gave me so much more when he introduced me to my wife. And I thank all my squadron mates in VA-72 and VFA-132 for the fun we had flying dangerous missions. Especially, Andy Ingram and Jim Tucker for helping me on some dangerous missions outside the cockpit when we scaled scary heights.

Capt. J. Craig Stencil, US Navy (Ret.), taught me to be a professional test pilot—and how to stay alive on dangerous flights when bad stuff starts happening.

I owe everything in my NASA career to Mr. George W. S. Abbey, who believed in me before he ever met me. I'm still learning from him.

I thank every crewmember with whom I worked not on this planet—Bonnie Dunbar, Marsha Ivins, Dave Low, Tammy Jernigan, Steve MacLean, Bill Shepherd, Lacy Veach, Mike Foale, Bernard Harris, Volodya Titov, Janice Voss, Jean-Loup Chrétien, Wendy Lawrence, Scott Parazynski, Dave Wolf, Anatoly Solovyev, Pavel Vinogradov, Paul Richards, Andy Thomas, Susan Helms, Yury Usachev, Jim Voss, Yuri Gidzenko, Sergei Krikalev, John Herrington, Mike López-Alegria, Ken Bowersox, Nik Budarin, Don Petit, Valery Korzun, Sergei Treshchov, and Peggy Whitson. Special thanks to my commander, Dan Brandenstein, and my pilots, Mike Baker, Eileen Collins, Mike Bloomfield, Jim "Vegas" Kelly, and Paul "Paco" Lockhart. I learned from every one of them.

I put my life in the hands of Dr. Phil Stepaniak and all the flight surgeons at NASA. I trusted Phil when it mattered, and he was right. Dave King, the center director, Marshall Space Flight Center, trusted me and shielded me from above.

The NASA community and I will never forget Special Agent Mike Sutton and the FBI for their efforts when we came together in Lufkin, Texas, in February 2003 to recover the human remains of the *Columbia* crew with dignity, honor, and reverence.

I thank every member of the Astronaut Office from the original Mercury Seven astronauts in Group 1 in 1959 through Group 19 in 2004, who directly and indirectly helped me. And I thank all the administrative personnel who supported me, especially Estella Gillette, Gloria Gibson, Mary Lopez, Elaine Kemp, and Lisa Navy.

To all my instructors, flight controllers, flight directors, flow managers, engineers, planners, processing technicians, support personnel, and everyone else who spent their life's energy to send humans into space—you made it all worthwhile and a giant blast! Jeff Bantle and Wayne Hale, in particular, made stellar decisions as my flight directors, in a role I believe is much harder than climbing on a rocket.

Prof. Debbie and Mike Griffin took loving care of my wife and daughters when I was occupied elsewhere in the solar system. I thank God for such great friends.

Donna Blankmann-Alexander will always be my mentor in all things safety culture.

Gary and Rita Considine are exemplars of integrity and friendship. Gary taught me much about leadership and doing and saying the right things. Rita shares my joy of unbridled happiness and unmitigated optimism, but she does it much better.

The mighty Max Weinberg taught me the ability to focus. He showed me the mental discipline required to achieve short- and long-term goals on stage and in life. He and Hal Blaine gave me music to take to the stars.

Capt. John W. Young, US Navy (Ret.), was an inspiration for decades, and especially when we worked together for five years on Mr. Abbey's staff while running human spaceflight operations. John is the best (and funniest) engineer and test pilot I ever met, on or off the planet.

Gen. Tom Stafford, US Air Force (Ret.), showed me how to make high-level programmatic decisions in service of our nation's space program.

Jim O'Brien mentored me and taught me gray areas are really black and white if you look closely enough. I thank all the auditors who helped me, especially Colin McIntosh, Tom Osorio, Barry Waddell, Andrew Springett, and Colin Anderson. I thank Robert Riley, John Putnam, and Dave Redeker for all the great conversations about leadership. Cheree Aspelin made unsolvable problems seem easy. I learned as much as I could from Steve Robinson and Tony Brock and became an honorary driller thanks to their expertise. And I owe my entire second career in the oil-and-gas industry to John Mogford and Mark Bly, who believed in me.

I cherish my late friendship and long e-mails with Jim Reason, the foremost author on safety, whose words of wisdom make this dark and dangerous world less scary.

I owe my latest career to Gary Terashita and Lisa Parnell, who provided awesome artistry in all matters literary. Also the incredible Keni Thomas—US Army Ranger, writer, and rascal (a real-life, guitar-playin', Black-Hawk-Downin', genuine, certified hero)—who brought us together.

And finally, the crews of the fatal *Apollo-1*, *Challenger* and *Columbia* missions are never far from my mind. Mike Smith taught me how to organize my training notes. Using his system, we created a one-thousand-page crew operations manual to fly the Space Shuttle. Judy Resnik selflessly devoted hours to explain the Shuttle's electrical system to me as a rookie. Laurel Clark was a kindred spirit. Even when she was prime crew, she always took time out of her hectic schedule to engage me in friendly conversation. I knew Willie McCool was the coolest pilot there ever was since I first met him. And Commander Rick Husband motivated me to be like him before and after his death. I hope I can meet them sometime in the next life and say thank you.

APPENDICES

These appendices represent a collection of leadership approaches and methods that managers can use to influence and inspire operating excellence in specific situations, as indicated. I have chosen to include each approach in these appendices for two reasons. First, I have found that few organizations fully embrace these leadership methods. Second, many individual leaders in the organizations have a great appetite for these particular methods. Often, they ask me to supply written versions of the subject material I enjoy explaining.

Sage and experienced leaders who mentored me over the years developed most of the content of these methods. I have simply codified their creative thoughts and effective practices. The few parts I originated were based on theories and higher-level concepts I learned from observing and working with great leaders who were successful in high-hazard missions.

My hope is you will use—and improve—these methods to *Control Risk in Your Dangerous World* and take your team and organization to new heights.

Appendix A. Seven Leadership Principles

Leaders in any endeavor can benefit from observing and analyzing the behaviors and attitudes of leaders who are successful in hazardous endeavors. Operating leaders speak and act differently, especially when they and their teams face danger.

The following principles underpin how operating leaders lead people:

1. The essence of leadership is enabling, motivating, and inspiring people to perform better individually and accomplish more collectively with higher quality, in service of a mission or pursuit of a goal, than they would have without the leader's influence.

Operators are motivated to create value, contribute to society, and make a positive difference in the world. Through great leadership skills, leaders can inspire the team to increase their performance and collectively achieve higher-quality results than previously thought possible. Leaders can make the people and the place better by building enduring capability for the future and leaving their part of the organization more capable than when they arrived.

2. The art of leadership is learned through experiences and mentoring over an entire career.

Theories and principles of leadership can be taught in the classroom. But the art of leadership is learned on the job, through experiences, with good mentoring over the span of an entire career. Leaders sometimes become great leaders after they have successfully passed through a severe test of their skills under the most demanding circumstances. The organization's intent should be to build the capabilities in potential leaders and help them become great leaders before meeting their crucible so they will be prepared to succeed.

3. Great leadership requires effective decision making honed through judgment, experience, and values.

In hazardous industries, people operate in unforgiving environments with extreme pressures, temperatures, and forces that can challenge the design capabilities of the equipment being used. These challenging situations may require complex engineering solutions. How successfully and sustainably these solutions can be delivered will depend upon the quality of leaders and their ability to make effective decisions based on superior judgment, vast experience, and values.

I include two subprinciples I learned from a brilliant leader, US Army Colonel Bernard B. Banks, PhD, professor and department head of Behavioral Sciences and Leadership, United States Military Academy at West Point:

3A. *The personal values of a leader reveal themselves in adversity.* Character matters. You cannot overcome in minutes what was instilled over years. The character of a leader is built over time and nurtured with care and attention. Character must be built long before adversity is faced.

3B. *Intent governs behaviors in crisis.* Purpose matters. Great leaders help their workforce understand the purpose of the mission. With the right intent, based on the purpose of the mission, leaders and operators are more likely to behave correctly in crisis situations.

4. Safety is a value in service of the mission; *safety enables great performance, which creates mission success.*

People who work in hazardous environments must never forget the danger. Leaders must help their workforce work safely. The best leaders recognize safety does not compete with production in the long term. Operating safely and systematically enables the workers to maximize productivity and increase long-term value and viability of the company by preventing accidents and simultaneously accomplishing the most possible within the current constraints of operations, maintenance, engineering, regulations, condition of the equipment, and readiness of the people. Safety is good business.

Through creating a mind-set of continuous vigilance and searching for vulnerabilities, leaders can influence their workforce to identify hazards and risks, to anticipate the changing shape of risk, and to prevent accidents. Together, leaders and the workforce can help the organization become trusted to safely accomplish its mission.

Leaders improve *performance results* (outcomes) by using effective *leadership and culture* on the social / human side, and proper *systems and processes* on the technical / systems side (inputs), to influence their teams to conduct safe, compliant, and high-quality operations (operating excellence).

5. The organization is responsible for providing the development opportunities and career paths to create great leaders.

The long-term health and viability of the company depends upon creating leaders for the future in a sustainable supply at all levels in the organization. Leaders rely on the organization to provide development opportunities and career paths that support the accumulation of leadership skills through successful experiences with progressively greater responsibilities as they advance during their careers.

Career paths should be in place and visible to employees across discipline groups that build experiential learning vertically through increasing levels of responsibility, and horizontally through wider foundational experience, with leadership prerequisites in place between points of progression. With the proper opportunities and career paths from hire to retire, leaders will continue their legacy of contributing value to society long into the future.

6. Leadership and culture are as important as systems and processes.[34]

Leaders are expected to exemplify the organization's values at all times. The leaders can inspire and influence their teams to infuse the values into their work. With ever-deepening leadership skills, leaders will be able to tap into the power of people working together to use systems and processes in a human way within a great operating culture to create long-term value.

7. The success of the organization depends on a culture of *commitment* and *accountability* for operating excellence embedded at the top and middle levels of leaders.

Senior leaders set the tone in any organization. Leaders who are not performing in the right way should be moved quickly to jobs where they can be more effective. No apologies are necessary. The mission is too important. Opportunities will be created for leaders with the right knowledge, skills, and attitudes to be promoted quickly into top- and middle-level leader positions.

The goal of the leaders is to improve performance results (outcomes) consistently over the long term. But the operators—the front-line workers—are the people who create value for a company by accomplishing the mission while working in hazardous environments. The leaders create success by enabling their workforce to accomplish the mission in a better way. Leaders with the right behaviors must be in place to create this success. Great operating leaders demonstrate the right behaviors through mastery in three leadership domains:

1. *Commitment*: Pursue a noble cause or mission;
2. *Caring*: Deeply care for their people who contribute to that cause or mission;
3. *Competence*: Learn quickly, search for vulnerabilities, reduce and control risks, create immediate learning across their teams and others in the organization.

---

34 This is an idea that I first heard spoken by an engaging leader, Mr. Robert Riley, who was the president of the largest company in Trinidad and Tobago earlier in his career. I had many conversations with Robert about his perceptive and powerful leadership principle.

Appendix B. Operating Leadership Behaviors

This appendix lists behaviors that are the foundation of operating excellence and leadership. Leaders should be committed to demonstrating these behaviors, which help all employees create a safe working environment and enable the front-line workers to produce as much as possible, under the existing safety and regulatory requirements, and operational and ethical constraints. This helps the company to become trusted to anticipate and deliver the needs of customers and society.

Setting Expectations

- Set clear expectations for your personnel. Include technical and social expectations to adhere to:
  - o Espoused organizational values;
  - o Expectations for leaders;
  - o Codes of conduct; moral and ethical behaviors;
  - o Rules, policies, and procedures.

    Have personal discussions with two-way dialogue to verify complete understanding of these expectations and garner agreement.
- Develop and publish descriptions of roles and responsibilities for each level of personnel in your organization.
- Set and communicate clear expectations for contractors about work requirements, work-related behavior, and adhering to the organization's Safety Management System and Operating System.

Managing Performance through Proper Support

- Help your people succeed by managing their performance with support, monitoring of responsibilities, and routinely exercising a process of accountability.
  - o Help your employees feel committed and accountable. Know your direct reports are not really accountable until you exercise a process of accountability and they accept their accountability to deliver.
  - o Convince your front-line operators (or direct reports) of the value of conforming to standards and basic operating principles. This will encourage commitment to conform.
  - o Train, teach, mentor, counsel, and coach your front-line operators how to conform to standards, basic operating principles, and advanced operating practices.
  - o Recognize and show appreciation for good individual and team performance and conformance with policies, rules, and operating techniques.

o Manage nonconformance immediately, fairly, and effectively.
- Conduct performance conversations routinely with your direct reports to provide constructive feedback to improve continually.
- Manage performance problems promptly.
- Solicit, encourage, collect, and value input from your front-line operators and direct reports on what isn't working correctly (policies that are confusing, unusable, incorrect, inaccessible, etc.). Resolve those concerns or elevate for resolution.
- Help your direct reports self-assess their leadership skills. Encourage them to develop an improvement plan and work with them to implement.
- Teach, mentor, and coach your personnel so they will be successful in their jobs.
- Show you value your employees who desire to improve.
- Communicate priorities to your team members and ensure they are working on tasks consistent with those priorities.
- If your people are overloaded, help them with clear priorities to reduce psychological stress, increase safety, and enhance their ability to perform successfully.
- Show care for your workers.
- Give and be willing to receive emotional support.
- In addition to your equipment and systems, protect the integrity and availability of your people. Don't push them past their human "red-line" limits of personal performance.

Safety and Operations
- Devote frequent and regular attention to safety in meetings with staff.
- No matter how strong the current performance has been, develop and continually update a plan to improve safety performance.
- Assess the operational and emotional readiness of your personnel to be successful in their jobs every day. Test your personnel to verify they have the proper knowledge, skills, attitude, and proficiency to perform their tasks.
- Help your direct reports and front-line employees develop a foresight for risk and build competence in managing and controlling risks. Convince them of the value of mastering risk-mitigation skills. Encourage them to practice using those skills.
- Teach your workforce to *manage and control risk effectively* by providing relevant and specific guidance for them to:
  1. Skillfully *Identify Hazards* (requires focused conscious effort);
  2. Accurately *Assess Risk* (requires progressive experience);

3. Correctly *Implement Controls* to mitigate danger and eliminate or reduce exposure of hazards to an acceptable level; and

4. Expertly conduct assessments of their risk-management and risk-control processes.

- Involve your workforce and people with the greatest expertise from inside and outside the team when assessing risk.

- Understand the dangers of risk tolerance and the insidious nature of complacency. Combat these effectively with your workforce. Create a collective *respect for hazards* by reminding your people of the risks on the following three occasions (minimum):

  1. Periodic reminders (for example, during regularly-scheduled safety meetings);

  2. After an incident, close call (near miss), or other learning opportunity;

  3. Immediately after complacency is observed.

  This third occasion is a powerful time for preventing accidents. The best way to observe and combat complacency is to have supervisors spending time in the field with their workers. These leaders must be attuned to signs of complacency and risk tolerance and must take immediate corrective actions when these are observed.

- Convince your workforce of the value of operating their systems in adherence to applicable Control of Work, Process Safety, and unit procedures.

- Encourage the use of your incident reporting system.

- Implement a method to verify the permitting process is being conducted effectively. Ask insightful questions to ensure members of your workforce understand the permitting process. Assess their ability to identify hazards, assess risk, and implement controls to manage and control risk. Ensure they are conducting the permitting process effectively (taking sufficient time and using proper skills).

- Teach error wisdom and *error-mitigation techniques*. Know all humans make mistakes. Encourage your people, both individually and as a group, to analyze human errors. Use the analyses to help your people develop techniques to:

  o Reduce the likelihood of errors;

  o Prevent incipient errors before they occur or have consequences; and

  o Mitigate the consequences of errors after they occur.

- Develop the ability of your people and teams to *maintain mindful awareness of risk* by giving specific and relevant guidance to:

  o Search for vulnerabilities in systems, equipment, and people;

  o Maintain situational awareness as they work; and

  o Anticipate the changing shape of risk.

- Understand and mitigate the conditions that may cause decreased *risk awareness* in your people. These may be organizational, situational, systemic, personal, emotional, or cognitive conditions that may cause a decreased ability to identify and control hazards to mitigate risk.
- Communicate and reinforce that all employees and contractors have the authority and the duty to stop or correct any unsafe activity. As the leader, you must clearly define for your employees and contractors the level of exposure that is unsafe. Most importantly, show you value the actions of employees and contractors when they do stop a job.
- Maintain a robust housekeeping plan to demonstrate you value safety and operations.

Rules, Policies, Procedures, and Practices

- Ensure rules, policies, procedures, and practices are correct, published, accessible, consistent, and understandable, and reflect the organization's best wisdom on how to conduct a task or process.
- Ensure your people are sufficiently trained and have the right tools to follow the rules, policies, procedures, and practices.
- Ensure your people know and follow the processes for creating, modifying, issuing, reviewing, and deleting rules, policies, procedures, and practices.

Creating a Learning Environment

- Create methods for your organization to learn valuable lessons.
- Model the learning culture behavior that you expect in others.
- Identify and encourage behaviors that indicate learning is occurring in your team.
- Make discussing issues and concerns psychologically and culturally safe for your people.
- Recognize and reward good learning behaviors and application of learning.
- Focus on education and personal growth by developing and maintaining a personal development plan.
- Help the people on your team to create their learning and development plans and help them execute their plans.
- Give your people the time to learn.
- Encourage your people to ask questions.
- Learn, do, and teach. Help develop others by sharing what you know and learn from your and other's experiences.

- Set the expectation that your people identify improvement opportunities. Teach your people how to improve their working environment using a process of continuous improvement.
- Solicit, encourage, and value inputs from your people regarding issues or problems that they face. Show appreciation. Never react defensively to issues that are brought to you. If you can't resolve the issue, elevate the issue higher in the chain of command, commensurate with operational severity. If you don't agree with their opinion, try to understand fully their rationale. Help them elevate their issue to higher authority.
- Ensure your people report all incidents, close calls (near misses), and opportunities to learn valuable lessons. Ensure that safety and the way your business operates remain a primary and consistent focus of your attention.
- Develop a culture that has the highest chance of reporting all incidents, close calls (near misses), and opportunities to learn. Convince your people of the value of reporting. The reputations of your people should be enhanced when they report personal errors in your organization; they and others will know that reporting will help the organization become a learning organization.
- Use a formal process to ensure that actions will be taken on lessons learned from both inside and outside your area of responsibility.
- Conduct a fair and just disciplinary process that your employees understand.
- Ensure your people know the process for creating, modifying, issuing, reviewing, and deleting policies and procedures that may be incorrect or need improving.
- Decide how you are going to improve the operating culture every day, and then take actions.

Personal Integrity
- Follow up on assigned action items, meet deadlines, and verify accuracy of completing the action.
- Eradicate incivility everywhere in the workplace. Respect every person at all levels. Treat others with the highest integrity after listening empathetically to how they desire to be treated.
- Resolve personnel issues quickly and correctly, using Human Resources support if needed.
- Treat your workforce, including contractors, with respect at all times. All are valued members of your organization.
- Behave with the highest ethics.
- Be a positive influence for your people.

- If you identify a problem with another person or group, introspectively assess how your actions are contributing to the problem, and change your actions or attitude to mitigate the issue.
- Self-assess your leadership capabilities. Develop a plan to maintain and improve competency, and work with your manager to implement.

Teamwork

- To support your team, demonstrate you are following the leadership principles your organization specifies.
- Foster team spirit by recognizing and appreciating cooperation. Acknowledge and give credit to all who helped the team succeed, especially those who do so without consideration of personal success.
- Help your team succeed by understanding what the team needs to do its job well. Proactively prepare them with proper techniques for staying safe and performing their jobs well. Identify and remove barriers that may prevent them from doing their jobs effectively.
- Create a team that works toward a common goal of safe and high-quality operations by setting clear expectations and adhering to plans and rules.
- Involve your team members in planning work tasks.
- Teach your employees to help co-workers be safe and productive.
- Use a process for your experienced workers to pass along their valuable knowledge to less-experienced members of the team.
- Encourage your experienced workers to value the inputs and questions from their less-experienced co-workers, who may have a different way of thinking, perceiving, and acting.
- Be sure each individual knows how she or he contributes to safety and the primary goals of the team and the business.
- After setting clear performance expectations for safety and compliance, follow up to check for conformance demonstrated by every member of the team.
- Exercise a process of accountability, specifically by managing the performance of your team members in a positive and principled way—continually.
- Eliminate a culture of blame. Effectively identify true causes that will help prevent future occurrences of incidents, close calls (near misses), mistakes, nonconformances, or unintentional violations.
- Earn trust. Know that you must give respect before you can earn trust.
- Foster effective teamwork across all areas and departments in the organization.
- Advance the organization's mission. Demonstrate that you and your team follow and support the organization's values.

- Build a sense of community. Continue to be valued contributing members of your external community.

Communication and Decision Making

- Ensure your team members know you will listen to what is important to them. Help bring out the best in everyone by listening effectively and using two-way dialogue. Actively listen, noticing and responding to feelings and unstated concerns in addition to the content of the message.
- Solicit, encourage, and—most importantly—*value* diverse ways of thinking and potentially unfavorable points of view or dissenting opinions. These make the ultimate decision better.
- Make informed and effective decisions after listening to, and valuing, the inputs of your people.
- Implement a process for quickly disseminating critical safety information to all personnel who need it. Track the personnel who did not receive the information so you can reissue before they need the information.
- Foster *effective* communication and two-way dialogue with team members and key stakeholders.
- Collect and share important information with teams.
- Explain the rationale for all decisions, programs, processes, policies, or initiatives that affect your people. If you don't know the rationale, try to get it. If you can't get it, try to enlist others' help to explain the rationale to your people.
- Make your communications relevant to your people. When translating objectives or policies from higher levels in your organization, show your people how the objectives relate specifically to their jobs.
- When required, admit you don't know and try to find out.

Quality of Operations

- Demand operating excellence in yourself and others.
- Ensure shift reliefs are thorough and robust. The off-going individuals should not be relieved until the on-coming individuals feel prepared.
- Take pride in what you do by setting high expectations and performing with higher expectations every day.

Growing Your Business

- Create leaders to succeed you and your other leaders.
- Improve the availability and reliability of your business units and systems by engaging your workforce in continuous improvement methodology.

- Improve cycle times and effectiveness of processes by engaging your workforce in continuous improvement methodology.
- Determine what society needs and supply those needs consistent with your organization's purpose.

Appendix C. Creating Accountability and Commitment

I have found few managers who understand the concept of creating accountability *before* an accident. In typical organizations, accountability is usually synonymous with blame and is administered *after* an incident or accident. In lower performing organizations, accountability is often levied on the lowest-level employee involved in the proximal cause of the accident, as identified by the investigation team.

Even in organizations with good reputations, accountability is often misunderstood. While facilitating a seminar, I had an opportunity to discuss the concept of accountability with a projects team leader. I asked him whether he felt accountable to his immediate supervisor. He responded he was not accountable to his boss but rather to a separate manager who was the recipient of his project work. I asked him how often that manager, who was essentially his customer, called him. He said she only calls when there is a problem.

In that one sentence, the team leader illustrated the depth of the problem. Accountability was not being created or conducted before problems surfaced. When I began to probe the understanding the participants had of the concepts of responsibility and accountability, I found a common misunderstanding.

Most managers and operators understand responsibility correctly as a sense of duty—to *do* something, to take action. When I ask managers and operators to describe accountability, they almost always describe nothing more than a higher sense of duty, or greater responsibility, as they oversee a team conducting activities. This description misses the point of accountability.

Accountability is a process of providing justification or *answering* to someone. The word *accountability* comes from the business practice of accounting, and it is meant to convey "opening the books" or providing an account. The first nuance managers often miss is accountability involves no fewer than two people, one who is asking questions and one who is giving the account. Additionally, when this process is conducted *before* an accident, the senior participant can ask insightful questions and provide helpful guidance or proper resources to prevent the impending accident.

I learned this process of accountability from my boss and mentor, Mr. George Abbey, who led human spaceflight at NASA. Every two weeks, I was required to answer his questions about my responsibilities, or what I was doing, as director, Flight Crew Operations. Because I knew he would continue to ask me questions about my responsibilities, I felt accountable to him. This feeling of accountability increased my personal commitment to doing the right things as I fulfilled my responsibilities in support of the mission of human spaceflight.

Responsibility is a sense of duty to do something. Accountability is feeling that I must answer to someone. If no one is there to ask questions, then I feel accountable to no one. In this case, the only incentives I would have in fulfilling my responsibilities would be my personal value system and commitment to the mission. In a dangerous business, when the smallest errors can cascade to catastrophe, personal values and commitment, no matter how great, may be insufficient for preventing accidents. Sage advice, guidance, and sufficient resources given by a leader who is exercising the process of accountability may be the critical factors that influence the accountable personnel to prevent accidents and improve productivity for the organization over the long-term.

Through the years, managers frequently have asked me how to create a process of accountability. Here is what I have offered.

Accountability is a process that includes *genuine dialogue* (listed in *italics*) with your direct report, using the following five steps to reach a mutual understanding of the objectives:

1. Set clear expectations (for priorities, enacted values, and behaviors)
    What I want you to do, how I expect you to behave, and why this is important!
2. Set limits or boundaries of expectations
    What I don't want you to do and why not!
3. Provide sufficient resources
    Here is what I will provide. What do you need from me to be successful?
4. Establish a method of accounting and performance management
    How and how often I will review, discuss, and support your progress and assess the quality of your delivery, demonstration of behaviors, learning from the past, and predictions for the future of your activities.
5. Discuss impact of delivery (and consequences for nondelivery)
    Do you think these consequences are appropriate? Would you accept these consequences if you fail to deliver?
    5a. Confirm Commitment
    Will you commit? (Managing Leader should ask!)

Leadership Commitments

The following are typical commitments the front-line operators usually want to receive from their front-line leaders:

1. **Set clear expectations**—As a leader, I will tell my people exactly what I want them to do, what is required, and how their work will contribute to the organization's goals. I will use genuine two-way communication and true dialogue to verify their understanding and personal commitment.
2. **Create accountability**—I will understand how my people perform their work. I will apply fair and just consequences, both positive and negative as

appropriate, for the quality of work being performed. I will be honest and respectful always. I will never avoid difficult conversations.

3. **Ensure competence**—I will assess the readiness of my personnel continually. I will build the competence of my people. I will put my people in positions where they can use their skills effectively and will make no apologies if I must move someone to a different position where their skills would be more suitable.

4. **Ensure work is performed properly**—I will systematically verify the conduct and quality of work and the use of procedures at all levels in my chain of command to the front line. I will assess and improve my organizational system of conducting and controlling work. I will support a process to develop, review, and improve procedures so they merit being followed.

5. **Engage personnel in the field**—I will spend time in the field to listen to concerns and support my workforce. I will understand their concerns, remove obstacles, improve our system, and help my people be successful.

Appendix D. Policy Note—Astronaut Office Conduct and Performance

One of the most common requests I have heard from followers is to understand clearly what their leaders expect. Soon after I was assigned as director, Flight Crew Operations, for human spaceflight at NASA, I published the following policy note to all astronauts to set the proper expectations:

National Aeronautics and
Space Administration
Lyndon B. Johnson Space Center
2101 NASA Road 1
Houston, Texas 77058

Reply to Attn of: CA-00-40   *August 18, 2000*

TO: CB/All Astronauts
FROM: CA/Director, Flight Crew Operations
SUBJECT: Flight Crew Operations Directorate Policy

Astronaut Office Conduct and Performance

Safety
- Johnson Space Center (JSC) Safety Policy—The JSC Safety Policy is:
  1. All mishaps can be prevented.
  2. You must remove or control hazards at work.
  3. Management will help you maintain a safe workplace.
  4. Training employees to work safely is essential.
  5. Your continued employment depends on working safely and watching out for others.
  6. Working safely will result in the best possible performance.
- Voluntary Protection Program (VPP)—VPP (an OSHA program to reduce injury rates below industry average) gives the responsibility for safety to every employee.

Flight Assignments
- Flight Assignment Requirements—Flight assignment recommendations will be based on operational capability to perform the mission. The capabilities we look for in the crew will depend upon the mission requirements. In general, this assessment is based on (in no particular order):
  o Ability
  o Systems knowledge

- o Performance in simulators, T-38, Shuttle Training Aircraft, Neutral Buoyancy Lab
- o Performance during remote manipulator system and payload operations
- o Demonstrated aptitude for and desire to learn operational skills during training
- o Previous flight performance
- o Leadership capability
- o Judgment
- o Teamwork
- o Honesty, personal conduct, ethics
- Good performance is expected in the ground job and makes an astronaut eligible for flight assignment. An astronaut's personal conduct is expected to meet the highest standard and must not bring discredit to NASA or any organization with whom we interface. Astronauts will be removed from flight assignments, or consideration, if they do not perform the requirements listed above satisfactorily.
- We recognize that all technical assignments do not provide the same opportunity for *proficiency* in the above listed knowledge and skills areas. However, when opportunities for simulators and flying occur, all astronauts are expected to be fully prepared and to strive for self-improvement in operational skills. An aptitude for and desire to learn these necessary skills are more important than proficiency in the pre-assignment training period.
- Astronauts will not normally be notified that they are being considered for a flight ("penciled in"). Any personal situation affecting flight assignments, therefore, should be brought to the immediate attention of the Chief, Astronaut Office, so that FCOD may plan accordingly.

Principles of Operational Management

Our operational management philosophy is:

1. The Commander is responsible for the development of optimum Crew Resource Management (CRM).
2. The Commander must use good leadership skills.
3. All crewmembers must use good operational techniques to enhance mission success.
4. All crewmembers must operate as part of a team.
5. Crewmembers must understand the flight and payload systems.
6. The crew must foster team support from the Mission Control Center (MCC).
7. Crewmembers should communicate effectively with MCC.

8. The crew must solicit constructive criticism from the Instructor Team.

9. Crewmembers should ensure methods exist for identification and rectification of errors or omissions.

10. The crew should have a good method for handling dissenting opinions.

11. At least one crewmember should have situational awareness at all times.

12. The crew should have good and efficient techniques for recovering from breakdowns in operational management and situational awareness.

13. All crewmembers must have a method to recognize fatigue and other stresses in themselves and their crewmates, and have a plan to cope appropriately with these problems during operations.

14. The crew must have an execution plan for all critical or complicated operations during flight.

T-38

- Crew Operations—Crewmembers must be able to:
  o Control and fly the airplane precisely,
  o Conduct checklists flawlessly,
  o Communicate efficiently and effectively with other crewmembers,
  o Transmit and receive operational information with ground controllers without error and recall critical aircraft limitations and boldface procedures.
- T-38 Errors—Since we are humans, we all make errors. We must use good techniques and cockpit resource management to prevent or minimize errors. Mental discipline is required. You must have the ability to multitask. Errors in the cockpit can result in aircraft damage and serious injury to crewmembers or ground personnel.
- Spaceflight Readiness Training—The T-38 provides an invaluable training tool for spaceflight. The skills necessary to fly the T-38 safely and well are the same skills needed to fly in space. The T-38 allows crewmembers to develop these skills and maintain proficiency through practice. Conduct Space Flight Readiness Training on every flight.

International Space Station (ISS) Support

- ISS Support—Do not underestimate the difficulty of building and operating ISS. Support the work on software and hardware in the field to mitigate flight failures. Help develop effective procedures. Be aggressive to ensure tests are being properly conducted. Go to board meetings to help ensure the right decisions are being made. If they are not, elevate the concerns of the Astronaut Office. Be vocal. Be polite. Be relentless.

- <u>ISS Contingency Planning</u>—Once in space, modules and trusses may not fit together, connector pins may bend, display software may not operate correctly, and internal hardware may not function as designed. Therefore, ensure that there is a well-thought-out contingency plan that addresses these potential situations before launch.

Expedition Corps
- <u>Expedition Corps Branch</u>—The branch consists of the Astronaut Office's long-duration crewmembers. This branch has a group to take care of crew-support issues for the long-duration crewmembers. The organization of this support is currently being expanded to include tighter cooperation with the Space and Life Sciences Directorate.
- <u>Long Duration Missions</u>—These missions will be international missions. Training will be conducted not only in the US but also in Russia, Canada, and other countries. ISS crewmembers from the US will have to speak Russian in order to properly communicate with MCC-M, Russian crewmembers, and Russian trainers. ISS crewmembers must be flexible and adaptable due to changes in flight manifest and crew rotation plans. Each ISS crew will have a dedicated backup crew. Due to expected limitations in training facility time, backup crew training may require some innovation. Astronauts who do not want to be considered for a long-duration flight because of the associated hardships might not be considered for short-duration missions.
- <u>Debriefing Confidentiality</u>—Crew debriefings will be kept confidential. This is particularly pertinent to postflight debriefings. A confidentiality statement, drafted by the JSC Legal Office, has been signed by the Directors of FCOD, Mission Operations, the ISS Program Office, the Shuttle Program Office, the previous Phase I Office, and the JSC Deputy Director. This statement will be given to crewmembers prior to the debriefings. It specifies that the debriefing materials (transcripts, video, audiotapes, etc.) will not be given to anyone outside NASA.

Training
- <u>Instructor Astronauts (IA)</u>—An IA group has been formed within the Astronaut Office to pass on experience, teach operational techniques, enhance CRM, and help train and evaluate our astronauts.
- <u>Long-Duration Training</u>—The Expedition Crew Working Group is reworking the ISS training flow to make it appropriate for long-duration missions.

- Astronaut Recognition—Astronaut candidates (ASCANS) who successfully complete the basic ASCAN training will be awarded a silver astronaut pin and designated as either an Astronaut Pilot Candidate or Astronaut Mission Specialist Candidate. In the positions of Pilot Candidate and Mission Specialist Candidate, astronauts will continue to be trained and evaluated for spaceflight eligibility. When assigned to a flight crew, they will then be designated as Astronaut Pilot or Astronaut Mission Specialist.

Lessons Learned

- Errors—Everyone makes mistakes. What we encourage and endorse is that everyone should have a good plan to minimize mistakes and have good techniques to recover from errors that are made. Mistakes must be acknowledged and admitted immediately so that all can benefit.

Interacting with People Outside the Astronaut Office

- General—We must remember that we represent the entire astronaut corps whenever we interact with other organizations in meetings, speak with the press, speak at Public Relations events, or meet with elected officials. It is imperative that we be polite, project a positive attitude, and stick to only those things we are expert in when speaking in these forums. Be humble, be courteous, and be professional at all times.
- Meetings—It is never appropriate to blame another person or organization when issues don't seem to be resolving as you (we) would like, particularly in front of others. Never assume that because an issue appears to be handled poorly by another group that the people are not competent. Almost always, the group you might blame for preventing closure on an issue has done more homework than you have. Be a good listener, do your homework, then educate them on where their solution causes adverse impacts on the crew or mission. Always look for a way to compromise.
- Public Speaking—Before answering questions from the media, always think of a way to make your answer sound positive. Never speculate, and don't be afraid to answer that you don't know or don't have the necessary information. Be prepared for the unexpected by thinking ahead, such as anticipating what is appropriate to say on camera during live coverage of a launch scrub. It is nice to show a sense of humor; however, it is never acceptable to use humor that is not appropriate or may offend someone in the audience. Think about who will be listening, or who may read your comments later. It is always better to defer answering questions about policy to upper management than

to get caught unaware in a media ambush. Remember, stick to your areas of expertise.

Overall Priorities

We have a tremendous amount of work ahead of us, and many demands on our time. It is imperative that we all cooperate to get the job done. Help your office mates when you see them getting over-tasked. Be a volunteer. Keeping our overall priorities in mind helps each of us organize better. Our priorities in carrying out our day-to-day activities are straightforward. In order, they are:

1. Performing space flight and associated assigned crew activities
2. Direct mission support and technical job assignment(s)
3. Currency training (training received between assigned flights)
4. Professional and/or military development
5. Appearances (public speaking)

Summary
*   Be technically accurate
*   Be relentless in accomplishing your job
*   Be polite, friendly, and helpful
*   Be an outstanding role model, and be humble
*   Be honest and always have exemplary personal conduct

As Director, FCOD, I am ultimately responsible for the safe and successful conduct of all our ground and spaceflight operations. Our people are now deployed around the globe, and we will have astronauts continuously in space on the ISS. Even though we continue to improve our hardware, our activities still remain hazardous, and we must remain vigilant. When you are sitting on the launch pad with your crew, I must have confidence in your abilities to safely and successfully accomplish the mission. Your crewmates must be able to *trust you immediately, unequivocally, and completely*.

You must give me and your crewmates your very best—pre-flight, in-flight, and post-flight.

As Director, FCOD, I will give you my best to support you in your work.

We have the honor of working in a great program. We must justify the privilege.

Original signed by:
James D. Wetherbee, Captain USN

BIBLIOGRAPHY

Berry, Mick, MFA, and Michael R. Edelstein, PhD. *Stage Fright: 40 Stars Tell You How They Beat America's #1 Fear*. Tucson, AZ: See Sharp Press, 2009.

Chiles, James R. *Inviting Disaster: Lessons from the Edge of Technology*. New York, NY: HarperCollins Publishers Inc., 2001.

Dekker, Sidney. *Drift into Failure*. Burlington, VT: Ashgate Publishing Company, 2011.

———. *Why we need new accident models*; Technical Report 2005-02. Lund, Sweden: School of Aviation, Lund University, 2005.

———. *The Field Guide to Human Error Investigations*. Burlington, VT: Ashgate Publishing Company, 2002.

Directive Number: CSP 03-01-003, *Voluntary Protection Programs (VPP): Policies and Procedures Manual*. Washington, DC: US Department of Labor, Occupational Safety and Health Administration, 2008.

Dorner, Dietrich. *The Logic of Failure: Recognizing and Avoiding Error in Complex Situations*. New York, NY: Metropolitan Books, 1996.

"Final Report, on the accident on 1st June 2009 to the Airbus A330-203 registered F-GZCP operated by Air France flight AF 447 Rio de Janeiro – Paris." Bureau d'Enquêtes et d'Analyses pour la sécurité de l'aviationcivile; Ministèrede l'Écologie, du Développement durable, des Transports et du Logement.

Gawande, Atul. *The Checklist Manifesto*. New York, NY: Metropolitan Books, Henry Holt and Company, LLC, 2009.

Helmreich, Robert L., Ashleigh C. Merritt, and John A. Wilhelm, "The Evolution of Crew Resource Management Training in Commercial Aviation," *International Journal of Aviation Psychology* 9:1 (1999): 19–32.

Joint Commission on Accreditation of Healthcare Organizations, *Issues in Provision of Care, Treatment, and Services for Hospitals.* Oakbrook Terrace, IL: Joint Commission Resources, 2004.

Kohn, Linda T.; Corrigan, Janet M.; and Donaldson, Molla S., editors. *To Err Is Human: Building a Safer Health System,* Committee on Quality of Health Care in America, Institute of Medicine. Washington, DC: National Academy Press, 2000.

Kolditz, Thomas A. *In Extremis Leadership: Leading As If Your Life Depended On It.* San Francisco: John Wiley & Sons, Inc., 2007.

Leveson, Nancy G. "A Systems Approach to Safety Engineering," Massachusetts Institute of Technology; http://www.ipa.go.jp/files/000005320.pdf.

Loukopoulos, Loukia D., R. Key Dismukes, and Immanuel Barshi. *The Multitasking Myth—Handling Complexity in Real-World Operations.* Burlington, VT: Ashgate Publishing Company, 2009.

Maurice & Katia Krafft: Lives on Fire, television documentary. American Public Television, distributor. Europe Images International, producer. Rights, contract period April 1, 2011–March 31, 2013, expired (not in rights).

Perrow, Charles. *Normal Accidents: Living with High-Risk Technologies.* Princeton, NJ: Princeton University Press, 1999.

Pinker, Steven. *How the Mind Works.* New York: W. W. Norton & Company, 2009.

Rasmussen, J., and I. Svedung. *Proactive Risk Management in a Dynamic Society.* Karlstad, Sweden: Swedish Rescue Services Agency, 2000.

Reason, James. *Managing the Risks of Organizational Accidents.* Burlington, VT: Ashgate Publishing Company, 1997.

———. *Beyond the Organizational Accident: the Need for "Error Wisdom" on the Frontline.* Quality Safety Health Care, December 2004.

Report, *Columbia* Accident Investigation Board. Washington, DC, August 2003.

Report. Presidential Commission on the Space Shuttle *Challenger* Accident. Washington, DC, June 6, 1986.

Stepaniak, Philip C. *Loss of Signal.* Washington, DC: National Aeronautics and Space Administration, 2014.

Weick, Karl E., and Kathleen M. Sutcliffe. *Managing the Unexpected.* San Francisco: Jossey-Bass, 2001.

Wolman, Moshe, and Manor, Ruth. *Doctors' Errors and Mistakes of Medicine: Must Health Care Deteriorate?* Burke, VT: IOS Press, Inc., 2004.

INDEX